KEYGUIDE TO INFORMATION SOURCES IN

# Cartography

KEYGUIDE TO INFORMATION SOURCES IN

# Cartography

A. G. Hodgkiss and A. F. Tatham

MANSELL PUBLISHING LIMITED
London

First published 1986 by Mansell Publishing Limited
(A subsidiary of The H. W. Wilson Company)
6 All Saints Street, London N1 9RL, England

British Library Cataloguing in Publication Data
Hodgkiss, Alan G.
  Keyguide to information sources in cartography.
  —(Keyguides)
  1. Cartography—Information services
  2. Cartography—Bibliography
  I. Title   II. Tatham, A. F.
  526′.07    GA102.5

  ISBN 0–7201–1768–2


Printed in Great Britain by
Whitstable Litho Ltd., Whitstable, Kent

To June and Margaret

# Contents

# Introduction

Cartography was defined by the British Cartographic Society in 1964 as 'the art, science and technology of making maps together with their study as scientific documents and works of art. Maps may be regarded as including all types of maps, plans, charts and sections, three-dimensional models and globes, representing the earth or any heavenly body at any scale. Cartography is concerned with all stages of evaluation, compilation, design and draughting required to produce a new or revised map document from all forms of basic data. It also includes all stages in the reproduction of maps. It encompasses the study of maps, their historical evolution, methods of cartographic presentation and map use'.

The literature of cartography is vast and wide-ranging, for mapmaking is an activity that reaches out to everyone in one way or another and one which overlaps numerous other disciplines. History, economics, politics, sociology, geology, botany and biology are merely a few examples of fields in which maps play a significant role. In recent times education in cartography has taken major steps forward and not only is cartography an important part of courses in geography and geology in higher educational establishments but also there are a number of courses specifically aimed at university degrees in cartography. This educational activity has meant that each year sees a proliferation of books, papers, technical reports, proceedings of symposia, etc., and the primary aim of this *Keyguide* is to direct the reader, whether he be directly involved with cartography and seeking information on a particular technical matter or someone working in another discipline who is anxious to expand his knowledge, to the most appropriate source for his purpose.

It is hoped that the *Keyguide* will form a useful reference aid for curators of map collections, librarians, cartographers, students in those disciplines on which cartography has a significant bearing, and for all those who work with or use maps.

Part I is a narrative exploration of the major areas of enquiry, commencing with an outline history of mapmaking, followed in Chapter 2 by an account of the origins of cartographic information and its development and utilization from medieval times to the present. Chapter 3 looks at the various agencies that are involved in the production of maps and other cartographic materials and at some of the major map collections. In Chapter 4 the current range of literature is discussed and a section on the more important of the many bibliographies and carto-bibliographies is included. Not only has there been a great expansion of literature in recent times, but so too has there been a proliferation of technical workshops, summer schools, symposia, one-day courses and similar events. Keeping informed about the venue, times and contents of these events is not easy and in Chapter 5 the reader is given some guidance as to where such happenings are publicized and reported on. Chapter 6 is devoted to the important topic—for all users of maps—of 'map care', and is largely concerned with matters such as classification and cataloguing, storage and conservation, map acquisition and the services that are generally offered by the curator of a map collection.

Parts II and III are annotated bibliographies of reference sources, Part II being concerned with the history of cartography and Part III with the contemporary cartographic scene. Each part includes sections devoted to different countries and to different types of map, as well as references covering aspects of the mapmaking process: information gathering, information processing, map production and map use. Limitations of space mean that the choice of references is restricted to major books and papers, and the selection is to some extent a subjective one.

Part IV is a directory of the various agencies involved in the making, distribution or conservation of maps from which information on their products or services may be obtained.

Finally, there is an extensive index of names, titles, organizations, topics and geographical locations.

Every endeavour has been made to ensure that the *Keyguide* is as up-to-date as possible and that the information provided is correct at the time of writing, but the authors and publishers accept no responsibility for problems arising from inaccuracies or omissions.

PART I

Cartography and Maps: the Subject and
its Literature

# 1  The History of Cartography

## Ancient Times

Maps are so much an integral part of everyday life today that it may come as a surprise to learn that mapmaking, albeit of a very primitive kind, pre-dates written history. While there is little tangible evidence for such a statement it can be justified by looking at the corresponding activities of primitive peoples in recent times and drawing conclusions by analogy. Such activities include the making of simple maps on a varied range of materials—stone, wood, animal skins, and in sand, for example.

The most quoted examples of recent primitive cartography are the so-called 'stick charts' of the Marshall Islanders in the Pacific Ocean. Bagrow (1964) states that these remarkable charts are unparalleled in the history of cartography. They are constructed from lengths of palm fibre tied together so that they point in various directions to illustrate the direction of the wave fronts. Placed at the intersections of the lengths of fibre are shells which are meant to represent islands and it is in the representation of the wave patterns between islands that these charts are so important for inter-island navigation. Other recent examples of maps made by primitive societies include those drawn in sand by nomadic desert tribes and surprisingly accurate maps of trading routes carved in relief in wood by Eskimos to indicate the coastal features of Greenland.

In the fertile valleys of the Middle East many centuries ago, men faced similar problems and, with one recently discovered exception, the earliest extant maps are Babylonian. They take the form of tiny maps scratched with a reed on clay tablets and, despite their miniature form, they show that their makers had some understanding of the concept of scale. Babylonian maps may be classified into three groups: estate maps at relatively large scales; regional maps at intermediate scales; and subjective views of the world as their makers saw it at very small scales. The world was presented in disc shape with an encircling ocean. The earliest surviving Babylonian regional map is dated by Bagrow at *c.* 3800 BC and the earliest world map at *c.* 500 BC. Bagrow's *History of cartography* (1964) illustrates a number of these early Babylonian maps and also the cartography of primitive peoples in modern times.

While historians generally see the Middle East as the birthplace of mapmaking, it is now thought that the oldest extant piece of cartography is a pictogram of a Neolithic village carved on a rock in the Val Camonica in northern Italy. This recently discovered graphic portrayal of the stilt houses, fields and village paths is believed to have been made *c.* 5000 BC and it vividly illustrates that even in such far-off days men had some idea of how to depict the spatial organization of their immediate environment.

Tangible evidence of mapmaking in ancient Egypt is virtually non-existent though it is known that surveying techniques of a rudimentary kind using measuring rods were employed before 3000 BC and that maps were made in order to redefine property boundaries when markers were washed away in the periodic inundations of the Nile. Such maps, made to record land ownership and the extent of property holdings, formed a basis for tax assessment and are known as *cadastral* maps.

In China the earliest extant map (fourth century BC) is a planning map while less than two centuries later the Chinese produced sophisticated topographical and military maps (Hsu, 1978, 1983).

Considerable progress was made by Greek scholars, who inherited some of the mathematical and cosmographical ideas of Babylonia. Initially Greek philosophers made similar schematic world maps in which the Earth is seen as a flat circle surrounded by water. During the fourth century BC, however, Greek scholars speculated on the possibility of the Earth's being spherical in shape, a revolutionary theory based to some extent on scientific observation and partly on the concept that the Earth, as the master creation of the gods, must necessarily be the perfect shape—spherical. Acceptance of this theory naturally led to speculation about the Earth's dimensions and Eratosthenes made remarkably accurate estimations of its circumference. Regrettably, however, the underestimations—7,000 miles too low—of another philosopher, Posidonius, were generally adopted and these erroneous figures were to have significant repercussions in future mapmaking.

Without the aid of some form of reference system it is difficult to locate places on a map and Eratosthenes pioneered the use of a reference grid composed of a framework of irregularly spaced parallels and meridians. In so doing he paved the way to portraying the spherical shape of the earth on a flat surface—perhaps the most difficult problem to be faced by cartographers and other scientists, who have devised innumerable map projections in an attempt to find the most nearly perfect solution. Early scholars faced another serious difficulty in that the very limited area of the known world or *oekumene* had somehow to be fitted into the vast mass of the sphere. One attempt, by Crates of Mallos, was to construct a globe in which four large balancing land masses were separated by narrow stretches of water.

Unquestionably the greatest contribution to Greek cartography was made by Claudius Ptolemy of Alexandria. In his work *Geographia*, Ptolemy provided all the necessary ingredients for mapmaking. Prepared *c.* AD 150, *Geographia* included a discussion on mathematical geography and cartographical theory, a discourse on map projections (of which Ptolemy was a pioneer) and a list of 8,000 place names with their degrees of latitude and longitude. Although the manuscript text was accompanied by maps it is not known whether Ptolemy himself prepared them. Some serious defects marred the impact of Ptolemy's work, largely because of his acceptance of the erroneous calculations of Posidonius. For example, he estimated that in an east–west direction Europe and Asia together occupied half the circumference of the globe instead of the true 130°; he elongated the Mediterranean Sea, omitted the Indian peninsula and showed a great river flowing across the Sahara. Despite these misleading errors, which were repeated by numerous mapmakers centuries later, it was now possible to map the world on a scientific basis rather than the speculations of contemporary philosophers.

With such scientific material at the disposal of would-be mapmakers it might seem that the scene was set for an era of steady progress. Nothing could have been further from the truth and indeed, throughout cartographic history, progress has been spasmodic; periods of relative stagnation are followed by a sudden upsurge of activity stimulated by theoretical, technical or geographical discoveries.

The Romans might have been expected to make outstanding contributions to cartographic development in the widest sense. Instead, their preoccupation with administering their widespread empire confined their mapmaking activities mainly to essentially practical projects for military and administrative purposes. They had little time or inclination to speculate about a much wider environment. One very interesting piece of Roman mapwork does survive, however, albeit a twelfth-century copy of a fourth-century map. This is the so-called *Peutinger table* (after the German humanist, Konrad Peutinger, who owned it), a route map made to delineate the highways of the Roman Empire and to locate the garrison towns, watering places, temples and the like. In addition to such human items the map also showed natural features, such as rivers and mountains, that represented obstacles or aids to communications. The most striking feature of the map is its shape, for it was prepared on a scroll some 20 ft × 1 ft (6½ m × 30 cm), a format which meant that gross distortions were inevitable. Nevertheless it was a practical attempt to illustrate graphically the spatial disposition of important features within the Roman Empire and, like the cadastral maps of Ancient Egypt, it is a good example of special-purpose mapping. As far as Roman thinking about the world as a whole was concerned, the old idea reappeared of a flat disc surrounded by water. The most famous Roman world map was

Agrippa's *Orbis terrarum*, made *c.* 12 BC. Known now only from descriptions in contemporary literature, it was apparently circular in shape and incorporated detail drawn from Roman road itineraries. In later Roman times small world maps were made in which the known continents were placed symmetrically within the disc: Africa in the bottom right; Europe bottom left; Asia straddling across the top; the whole encircled by oceans. This simple arrangement was carried on to become the model for Christian mapmakers in the Middle Ages.

## The Medieval Period

Two diametrically opposed approaches were apparent in the mapmaking of medieval times. The one, theological, reflected the decline in scientific thinking consequent upon the collapse of the Roman Empire; the other, an essentially practical, down-to-earth approach by Genoese and Catalan mapmakers, produced exceptionally fine, accurate regional maps.

The theological outlook is seen in the so-called T–O maps, usually drawn as illustrations to monastic texts and of Roman derivation. Their construction, in which the encircling ocean forms the 'O', the Mediterranean the upright stroke of the 'T' and a line joining the rivers Don and Nile the horizontal bar of the 'T', is explained in the writings of Isidore, a seventh-century bishop of Seville: 'The earth (orbis) was named from its roundness ... Europe and Africa occupy half the world ... Asia alone occupies the other half. The former were made in two parts because the great sea (called the Mediterranean) enters from the ocean between them and cuts them apart.' Although maps drawn on the T–O scheme were generally small book illustrations, the principle was sometimes expanded and modified in great thirteenth-century world maps or *mappae mundi* such as the 13 ft (4 m) diameter Ebstorf map, which was regrettably destroyed during the Second World War, and the extant Hereford World Map, a unique cartographic treasure that is on permanent public display in Hereford Cathedral. The Hereford map was made *c.* 1290 by Richard of Haldingham, a prebendary of Hereford Cathedral. Five feet (1.6 m) in diameter, it combines the Roman and medieval theological traditions by placing east at the top in the basic T–O form with Jerusalem at the map's centre and filling in an encyclopedic wealth of pictorial information culled from histories and bestiaries. Towns are portrayed as small groups of buildings; rivers and mountain ranges are prominently shown and there is a profusion of mythical creatures and strangely deformed people. The map had little practical value except as a means of communicating medieval theological beliefs to the large numbers of people who visited Hereford and its cathedral.

The accurately constructed sea charts, or 'portolan charts' as they were known, made by the Genoese and Catalan mapmakers were in striking contrast to the odd mixture of fact and fantasy perpetrated in maps such as the Hereford Map. At last a major step forward in cartography had been made and the use of these charts, allied to the magnetic compass (which was introduced into the Mediterranean region during the thirteenth century), meant that mariners could navigate well away from the shores with a new confidence. The portolan charts, prepared in glowing colours on sheepskin, faithfully delineated the Mediterranean and Black Sea coasts, the emphasis naturally being on coastal detail—harbours, cliffs, rocks, etc.—rather than on the land areas, for these were practical charts made essentially for navigation.

The earliest surviving example of a portolan chart is the *Carte pisane* or Pisan chart dating from the late thirteenth century but Abraham Cresques, a Jewish cosmographer working in Majorca, produced his *Catalan atlas* in 1375, a magnificent pictorial world chart that extended the scope of the standard portolan chart to take in the whole of Asia. In so doing Cresques used an amalgam of data taken from contemporary nautical sources and from the narratives of thirteenth- and fourteenth-century travellers in Asia—notably Marco Polo.

Outside the confines of Europe there was an advanced school of Islamic cartography, for in the ninth to eleventh centuries Arab scientists were superior in knowledge and technical skills to their opposite numbers in Europe. The most outstanding examples of Islamic geography came from the widely travelled and scholarly Idrisi. The Norman king Roger II Guiscard of Sicily (1097–1154), an enthusiastic patron of science and learning, commissioned Idrisi to prepare a volume that would bring together all available information on places with their latitude and longitude, their distances apart and their position in relation to climatic zones. After fifteen years of research a book and map were prepared but the map, which was prepared on a large silver tablet, was destroyed in 1160. In 1154, however, manuscripts of the book were produced in both Latin and Arabic and a detailed map drawn up on seventy sheets was accompanied by a small circular world map. Idrisi's work was superior in both concept and content to the reactionary theological cartography of medieval Europe. However, European cartography was soon to come to life.

## The Cartography of the European Renaissance

Several mapmakers, notably Fra Mauro, a monk of Murano (an island in the Venetian lagoon), followed the example of Cresques in producing maps which extended the range of the standard nautical chart. Fra Mauro's world map, made in 1459, is a circular *tour de force* that represents the zenith of medieval theological cartography. The circular format, drawn with south at the top, meant that distortions were unavoidable, but the map presents a vast amount of detail, albeit largely pictorial, and including many roads not so far shown on maps, the data for which were clearly gained from the itineraries of contemporary travellers. Now, however, at last there came a breakthrough in European cartography, and a new perspective was added to the entire mapmaking process. This breakthrough, one of the major landmarks in cartographic development, came about as a result of three significant developments.

The first contributory factor to a cartographical renaissance came with the reintroduction into Europe of a manuscript of Ptolemy's *Geographia*. While manuscripts had been preserved by the Arabs, *Geographia* had not been available to European scholars for centuries. Its repossession meant that scholars quickly translated the text into Latin and geographers soon got down to the task of redrawing the Ptolemaic maps. With the newly discovered craft of printing came a spate of printed editions of Ptolemy, creating an ironical situation in which the most up-to-date technology was being used to spread a view of the environment as seen by Ptolemy many centuries earlier and which recorded his errors for a long time to come.

Secondly, the geographical discoveries of the great navigators who were probing routes to the Indies in both easterly and westerly directions brought in a great deal of new geographical data that had to be incorporated into contemporary maps.

Finally, the application of printing from engraved surfaces—initially from wood blocks, later from copper plates—meant that maps could now reach an audience wider than a handful of scholars, statesmen and wealthy merchants. Multiple impressions or prints could now be taken from an original engraving, each being a faithful reproduction of the original. In this context it should be remembered that when we speak of 'original' maps today we are normally thinking of an original impression rather than one single unique piece of artwork.

The world maps of the Renaissance period tended to be highly pictorial in presentation for it was natural that exciting new discoveries of flora and fauna in distant lands would be of great interest to the public and should be incorporated into maps. Some of this portrayal of animals, birds, fish and sea creatures had been derided as mere space-filling to cover gaps in knowledge but the zoologist Wilma George (1969) argues convincingly that the mapmakers knew what they were about and that Renaissance and later maps provide much useful source material on contemporary natural history.

In their geographical content material gleaned from the new discoveries was being added to world maps. For instance manuscript maps by Juan de la Cosa (1500) and Cantino (1502) showed part of the American continent. The first engraved map to show contemporary discoveries about this new continent was that of Giovanni Contarini (1506), but a superb twelve-sheet woodcut map made by the Alsatian geographer Martin Waldseemüller in 1507 was the first to show America as a continuous land mass from north to south and was also the first to use the name 'America', which Waldseemüller placed on the southern part of the continent.

In 1522 the survivors of Magellan's scheme to circumnavigate the globe returned to Seville and the data they brought with them meant that further radical alterations were necessary to bring Ptolemy's world picture up to date. America could now be displayed with greater accuracy and the enormous extent of the globe occupied by the Pacific Ocean was now recognized. One of the earliest maps to illustrate this new outlook on the world was made by the

Portuguese cosmographer to the King of Spain, Diego Ribeiro. Not only was Ribeiro's map (1529) remarkably accurate in its geographical portrayals but in its quality of execution, the excellence of the calligraphy and the artistry of its presentation it represents a supreme example of graphic communication.

Many printed editions of Ptolemy's *Geographia* continued to be published and the history of these editions is traced by Stevens (1908). Ptolemy's scheme of twenty-six regional maps and a world map formed the prototype for ensuing geographical atlases. Facsimile reproductions of all twenty-seven maps from a superb copper-engraved edition produced in Rome in 1490 are included in the famous *Facsimile atlas* (1889) of A. E. Nordenskiöld. During the Renaissance period, however, it became increasingly obvious that the original Ptolemy maps failed to meet the requirements of contemporary users for they did not provide an up-to-date world view. The solution was to introduce *Tabulae modernae* or 'modern maps' to supplement those of Ptolemy. In his magnificent woodcut edition of 1482 Nicolaus Germanus added five new maps and by 1540 Sebastian Münster had increased the total number of maps to forty-eight.

Like other branches of art, mapmaking has fallen into national schools and in the late fifteenth and early sixteenth centuries the centre of geographical studies was southern Germany. Here too, the craft of wood engraving prospered and encyclopedic works known as *Cosmographiae*, which discoursed on geography, astronomy, history and the natural sciences, were produced by this process. They were normally copiously illustrated by woodcut maps, views and portraits. Petrus Apianus, or Peter Apian, professor of mathematics at Ingolstadt, produced a highly successful *Cosmographia* which he conceived as a teaching manual and which included maps illustrating Apian's views of world geography.

Outside southern Germany excellent cartography was being produced elsewhere in Europe, notably in Italy where the map trade was centred on Rome and Venice. The leading Italian cartographer, Giacomo Gastaldi, had a prolific output of finely engraved maps including regional maps of Italian provinces. Italian mapsellers such as Antonio Lafreri (1512–77) were regularly assembling to customers' orders collections of maps made up from whatever they had in stock. Sixty or seventy such atlases survive and they have been termed Lafreri atlases.

## The Low Countries: Ortelius and Mercator

In the late sixteenth century cartographic supremacy was wrested from Italy and Germany by Antwerp and Amsterdam and a new 'golden age' of mapmaking commenced. The Flemish geographers Abraham Ortelius (1527–98) and Gerardus Mercator (1512–94) saw the advantages of producing volumes of maps that would represent the summation of regional cartography in a reasonably handy format. Unlike Lafreri, in whose atlases maps could be a variety of sizes, Ortelius insisted on uniformity throughout his atlas, which was published in 1570 under the title *Theatrum orbis terrarum*. The seventy maps of *Theatrum* were gathered from the most reliable sources; Ortelius was a scholar and an editor rather than an original mapmaker, unlike his even more famous contemporary, Mercator. The latter's *Atlas* (the first use of the word) was an extremely ambitious affair and was published in several volumes, the last one appearing only in 1595, a year after Mercator's death. Mercator was a multi-talented man who combined skills in land surveying, instrument making, calligraphy, engraving and cartography. He was responsible for the revolutionary projection that bears his name and which has proved so beneficial to navigation. Mercator's influence during the latter half of the sixteenth century on geographical thought and on the construction and design of maps grew until it dominated the whole of Europe (Skelton, 1969).

After Mercator's death the atlas became the dominant cartographic form, increasingly more magnificent volumes being produced by the great Dutch publishing houses of Blaeu and Jansson, and their successors Visscher, De Wit, Valck and Schenk. Many of these atlases made little significant contribution to cartographic progress as far as content was concerned but in clarity of layout, elegance of design, quality of execution and fine engraving, the Dutch publishers did a great deal to improve the map as a means of graphic communication.

## Great Britain

So far Great Britain had hardly been in the forefront of mapmaking though there had been isolated instances of maps that were landmarks in cartographic development: the already mentioned Hereford *mappa mundi*; four maps of Great Britain made by the monk of St. Albans, Matthew Paris (*c.* 1250); and in particular, the extraordinary road map of Great Britain (*c.* 1350) that is now preserved in the Bodleian Library, Oxford and is known as the Bodleian Map or the Gough Map. This map was unique in its day for its amazingly accurate delineation of the road network of Great Britain, for its indication of towns and the distances between them, and for the enormous improvement it showed in the coastal outline of the country. The Gough map was reproduced in facsimile by the Royal Geographical Society in 1958 and analysed by E. J. S. Parsons in an accompanying memoir that lists all the place names and their modern form.

For a long time Great Britain continued to be mapped as a single unit, a notable example being the first copper-engraved map of the British Isles (1546) by George Lily, a Catholic refugee living in Rome. From the late sixteenth century, however, the county became the basic unit for mapwork until the establishment of the Ordnance Survey and the first one inch to one mile maps at the beginning of the nineteenth century.

By 1570 a number of Flemish refugee craftsmen had introduced the art of copper engraving into England; surveying instruments and techniques were steadily improving and numerous treatises on the art of surveying, such as the *Pantometria* (1571) of Leonard Digges, were being published. Digges described an instrument of his own invention which in principle resembled a modern theodolite and which could be used to measure both horizontal and vertical angles. The landed gentry who came into estates as a consequence of the confiscation and redistribution of monastic lands needed up-to-date estate maps, a need that led to the emergence of a new body of professional surveyors known as 'land-meaters'. From this body emerged one of the great figures of British cartography, Christopher Saxton, a young estate surveyor who had the good fortune to be commissioned by a wealthy Suffolk gentleman, Thomas Seckford, to survey and map all the counties of England and Wales. Little is known of the surveying methods used by Saxton, but Evans and Lawrence (1979) note that several instruments were available to him and that the inception of triangulation towards the close of the sixteenth century implies that he must have sought out elevated positions from which to view the territory he wanted to map. (In works which will be of great interest to modern professional surveyors and cartographers, Richeson (1966) presented an excellent account of the development of land surveying in England, and Thompson (1968) described the growth of surveying as a profession.) Saxton completed his task in a remarkably short space of time, the survey taking only seven years, and in 1579 he published his *Atlas of England and Wales*, the first national atlas to be published anywhere. Many of the maps were engraved by Flemish refugees but Saxton also employed English craftsmen to good effect. On his maps he made good use of conventional symbols: different-sized hummocks to represent hills, a graded series of symbols for settlements, a pale fence round the great estates, clustered trees for woodland and forest. He stressed the accurate placing of towns and emphasized those features which would aid or prove a barrier to movement, such as bridges over rivers, marshes or ranges of hills. Roads were omitted, for some reason. Study of the Saxton maps is facilitated today by the series of facsimile reproductions available from the British Library.

John Norden (1548–1626) was also an estate surveyor of proven ability but, unlike Saxton, he lacked patronage and was beset by financial problems so that his scheme for a series of county histories accompanied by maps failed when only five counties had been completed. Norden's were the first county maps to show roads and he also introduced the triangular distance table into Britain, a device still used today, by which distances between towns can be read off easily and quickly.

The 1607 edition of William Camden's *Britannia* included county maps derived from Saxton and Norden but from the early seventeenth century British mapmaking began to be dominated by London-based publishers. John Sudbury and George Humble launched the most popular cartographic venture of the day, John Speed's *The theatre of the Empire of Great Britaine*, a volume designed to accompany Speed's *History of Great Britaine*. Remarkably, the latter is now almost completely forgotten while the maps grow ever more popular. Speed, like Ortelius, was no originator but an editor/compiler who assembled his material in study and library, preparing map layouts for his splendid engraver, the Dutchman Jodocus Hondius. Speed's major contribution to cartographic development was the inclusion of a town plan, usually of the county town, on each map and taken together these form the earliest collection of engraved English and Welsh town plans. Speed also introduced hundred boundaries but, like Saxton, strangely omitted roads.

Although Britain was now mapped in considerable detail there was no ensuing era of steady progress. Successive mapmakers plagiarized the works of Saxton, Speed and Norden, rather than expending time and money on making their own surveys, and little progress was made until 1675 when John Ogilby published *Britannia*, a folio volume in which the principal roads of Great Britain were presented in strip form on 100 plates with six or seven strips per plate. Ogilby's work was progressive in various ways: it was based on the first really systematic survey of British roads; it introduced the statute mile of 1,760 yards; it established the one inch to one mile scale as a standard. Its influence was such that as far as road books were concerned, only derivatives were produced until John Cary made a superb new road survey a century later. The publication of *Britannia* also meant that thenceforward roads could be included on all series of county maps. Indeed the second half of the eighteenth century became a propitious time for British cartography with the inception of an award by the Society for the Encouragement of Arts, Manufactures and Commerce (later the Royal Society of Arts) of £100 for an original county survey at the one inch to one mile scale. This offer resulted in a number of high-quality county maps, the first winner of the award being Benjamin Donn for his map of Devon (1765).

## North America

At this time too European mapmakers were finding a fruitful field of operation in North America, where the organization of the colonies and the spread of settlement made the production of good, accurate regional maps imperative. Several Frenchmen, among them Guillaume de l'Isle, made important contributions to the mapping of North America and Dutch and English mapmakers also played their part. The peak of colonial cartography is Dr. John Mitchell's 1755 *Map of the British and French dominions in North America*, a map that had a unique place in American history for it was used at the Peace Conference of Paris in 1783 to lay down the boundaries of the United States. Military maps were produced in great numbers during the Revolutionary war and many are superbly reproduced in the *Atlas of the American Revolution* (1974). After the war the task of subdividing and settling the great tracts of land inherited by the new republic had to be faced. Thomas Hutchins, as Geographer of the United States, was entrusted with the task of carrying out a major cadastral survey and by the time of the American Civil War (1861–65) a large amount of territory had been surveyed. At the end of the Civil War maps were required for a variety of legal and administrative purposes and a profusion of cartographic material appeared, notably the American county atlas, a form of publication that continued to be produced to the beginning of the twentieth century.

# The Beginnings of Modern Cartography: National Surveys

The eighteenth century is regarded as a time of cartographic reformation. Constant warfare created such a demand for up-to-date and accurate maps that private cartography could no longer cope. From 1750 nations began to organize surveys on a national scale, beginning with *Carte du Cassini*, a survey which covered the whole of France in 182 sheets at a scale of 1:86,400. This new conception soon had repercussions across the English Channel and in 1791 Britain established the Trigonometrical Survey (later the Ordnance Survey). A national survey of Britain was begun and the first sheet resulting from this survey appeared in 1801 at the one inch to one mile scale. This particular sheet was a map of Kent privately published by William Faden; the first numbered sheet of the national series did not appear until 1805. For obvious military reasons the earliest sheets to be published were of southeastern England and progress in the south was fairly rapid. The series of 110 sheets was not completed, however, until 1873 with the publication of the Isle of Man sheet (sheet 100). For study purposes facsimile sets of the early printings of the series (called the Old Series, sometimes the First Edition) are now published by Harry Margary and of the later printings by David & Charles, the latter set including the important rail network.

It was not long before other European countries followed the example of France and Britain in setting up national surveys, while further afield the work of the Survey of India made the subcontinent the best-mapped large territory in the world. As mentioned already, colonial mapping developed rapidly in North America and provided much-needed employment for private mapmakers, who, at home, had to compete with the new national organizations.

## Thematic Mapmaking

Up to the beginning of the nineteenth century mapmakers had been mainly concerned with the completion of locational gaps in the world map but as this overall picture was gradually filled in there began a new concern with the nature of phenomena in different locations, followed by attempts to assess their quantity or value. Geographers also became interested in regional studies and examined specialized themes such as communications, disease or the distribution of population. Allied to data from census surveys cartographers now had a multiplicity of material which they could transform into special-purpose or thematic maps. Edmond Halley, the astronomer, had already used thematic maps to demonstrate his scientific theories, publishing a map of trade winds in 1686 and a map of compass variations in the Atlantic in 1701 in which he used isolines (a generic term for lines joining points of equal value). Thematic maps, often concerned with social conditions (an example being the dot distribution maps of Dr. John Snow (1855), in which he proved that cholera in a particular part of London occurred only among those who drank water from a particular pump), appeared in profusion during the nineteenth century.

In the late eighteenth century William Smith made numerous observations concerned with rock strata and arrived at two highly important conclusions: first, that strata always succeed one another in a defined order, and secondly, that fossils found in a particular stratum are peculiar to that stratum, the sequitur being that strata can be identified by an examination of the fossils they contain. Smith incorporated his findings into geological maps, using a specially prepared base map constructed by the cartographer John Cary. These renowned geology maps covering England and Wales with parts of Scotland were published by Cary in 1815. From this beginning geological mapping proceeded apace throughout the nineteenth century and by the latter half of the century most European countries possessed detailed geology maps.

Communications were of great importance during the nineteenth century and private mapmakers, who in general mapping had to compete with the official body, the Ordnance Survey, turned their attention to mapping the growing networks of roads, canals, railways and the stage-coach and packet-boat routes. Railways were included on the later printings of the Ordnance Survey's Old Series One Inch to One Mile map but perhaps the most remarkable nineteenth-century achievement of the official body came with the production of ultra-large-scale (1:500 and 1:528) urban plans. Unrivalled in quality then or since, these large sheets display a staggering amount of urban detail: individual lamp-posts and drainage grids, bay windows, every pillar and pew in church interiors. To the student of nineteenth-century urban geography or social conditions these plans are an indispensable tool.

## Technological Developments

The nineteenth century was one of the great boom periods of cartographical history. It was a time of rapid progress facilitated by radical changes in technology that affected both survey and map reproduction. The introduction of photography in a new reprographic process—photolithography—greatly expedited map reproduction and also enabled maps to be printed in full colour. Prior to the nineteenth century maps were printed in black and white only and where required were coloured individually by hand. As far as survey was concerned, greatly improved instruments and techniques, allied to the beginnings of photography from the air, were producing greater speed and accuracy of survey. Photography from tethered balloons gave the mapmaker a new perspective on the environment and it was a natural step that cameras should later be mounted in aircraft.

## Cartography Today

In the modern world maps have become an integral part of everyday life. As individuals we perhaps take our use of maps for granted and yet there can be few days when any one of us fails to make use of one form of a map or another, whether it be simple black-and-white maps in the daily press which inform us about current events; whether it be the weather map on the television screen; whether we plan a motoring trip, a walking tour, a cycle ride, a holiday, or even want to assess the environment of a place in which we propose to live. In local government maps are essential in drawing up all kinds of schemes: roadworks, sewerage, housing, parks and gardens. At the national level maps are an important tool in government strategic planning, whether for political, economic or military purposes. In education of course maps are indispensable; they play a vital role in the teaching of numerous disciplines at all levels: geography, history, economics, most of the social sciences and many of the earth and life sciences. The educational field is an area in which thematic maps as well as topographic maps are widely used and in the thematic area automated processes are being increasingly applied to the mapping of statistics, for the computer offers distinct advantages in the rapid processing of statistical data and transforming it into map form.

## The organization of contemporary cartography

Prior to the late eighteenth century mapmaking was generally the concern of private mapmakers, publishers and commercial establishments. Today in most countries the responsibility for the basic original trigonometrical survey rests with the particular national survey body: in the case of Britain the Ordnance Survey; in the United States the U.S. Geological Survey; in Switzerland, the Eidgenössische Landestopographie (National Survey) and so on. The Federal Republic of Germany is one country that departs from the usual pattern of centralized survey, for in Germany official cartography is mainly in the hands of land-survey administrations situated in each of the *Länder*, bodies that look after the large-scale mapping of their own *Land* together with a certain amount of small-scale work. A decentralization of this kind can lead to difficulties in standardization and uniformity, and to cope with this problem a working group composed of members from federal bodies and from each *Land* was inaugurated to keep a watching brief on output and to deal with the problems of preparing small-scale maps. In 1952 it was decided that the Institute for Applied Geodesy should prepare such maps. Australia, too, differs from the normal pattern. Federal survey is in the hands of the Director of National Mapping in the Department of National Development but separate states also have their own mapping organizations.

Generally speaking, then, original survey at a national level is the prerogative of central-government-backed bodies but by no means has the demise or even the decline of private mapmaking occurred. Indeed the private sector everywhere produces an immense and varied output of material, usually in the form of compiled maps constructed on official survey base material. Official bodies normally charge a copyright fee for allowing their data to be used in this way but in the United States maps are regarded as being in the public domain and free usage is permitted. Much of the commercial output is concerned with maps for educational and leisure needs: atlases, globes, wall maps; maps for motorists, cyclists, walkers, yachtsmen, canal boating enthusiasts, climbers and others. There is also considerable use of maps in commercial advertising but here we are in the field of the graphic artist rather than the cartographer. The use of maps in publicity can often be colourful, innovative, even amusing, for the designer need take little heed of topographical accuracy or cartographic convention.

European countries, and developed nations generally, have a large number of commercial mapmaking firms. For example, in Britain there are large firms such as the old-established Edinburgh establishment of John Bartholomew & Son Ltd., together with a number of newer and smaller companies. Cartographic production in Britain is located in just a few centres, notably Southampton, Oxford, Edinburgh, Taunton and London. Other important companies in Europe include Kummerley & Frey of Berne, Freytag-Berndt of Vienna and Innsbruck, Westermann Verlag of Braunschweig, Falk of Hamburg, Cappelen of Norway, Esselte of Sweden and many others. North (1982) provided a breakdown of maps currently being produced in the United States, where federal agencies distribute some 161 million maps per year while the private firm of Rand McNally distributed around 50 million maps to businesses and 25 million to private individuals. Another Chicago company, Denoyer-Geppert, distributed $2\frac{1}{2}$ million maps in 1981 so it is clear that the output of cartographic material is immense and that private cartography, far from being in the doldrums, seems to be very much alive.

The contribution to mapmaking of individual, free-lance mapmakers must not be overlooked. John Flower, for example, produces work of the highest quality in Britain, working to commissions from publishers and commercial advertisers; the panoramic maps, often of Alpine locations, by the artist Professor Heinrich Berann of Lans near Innsbruck, are world-famous and have inspired a number of imitators; in the United States the work of Richard Edes Harrison for *Fortune* magazine was innovative and exciting while the terrain maps of Erwin Raisz presented land use on a continental and regional scale with considerable impact; in Switzerland the work on terrain representation by the doyen of European cartographers, Dr. Eduard Imhof, has played a significant part in improving the quality of this aspect of mapmaking and in making Swiss maps among the finest examples of cartography in the world.

After the Second World War a new development in Britain was the setting up of small cartographic units in higher-educational establishments, first in universities, later in some polytechnics and colleges. The main emphasis was on small-scale thematic maps prepared for illustrative purposes, either for books and papers or as visual aids in academic teaching. Today similar units are to be found in universities throughout the world and the output of statistical mapwork is considerable.

## Careers in cartography

Today cartography with its combination of skills—survey, draughtsmanship, statistical expertise, editorial know-how, artistry, graphic design and reprographic skill—offers an interesting and challenging career. Employment is offered in a wide range of establishments: large official bodies such as the Ordnance Survey or the Admiralty Hydrographic Service and smaller official organizations such as the Forestry Commission and Institute of Terrestrial Ecology; private commercial firms; oil companies; and educational establishments. Free-lance work too is available.

Training for the profession was 'in house' for many years, the establishment of cartography as a separate discipline in departments of higher education being a fairly recent development. Chairs of cartography have been established in the United States, Canada and the Netherlands but Britain has still to install its first Professor of Cartography. While in-house training is still provided in many establishments, degree and diploma courses in cartography are now available in some universities and technical colleges. R. W. Anson prepared a booklet for the British Cartographic Society [779]* which sets out in detail the educational courses available and lists all the establishments in Britain which offer employment opportunities in cartography. For the individual who is unable to attend an organized cartographic course the prolific supply of textbooks and journals is helpful and so too are the annual symposia, technical workshops and summer schools organized by various cartographic societies.

## *References and Further Reading*

Anson, R. W. *Careers in cartography*. Oxford: British Cartographic Society. N.D. 19pp.

Bagrow, Leo. *History of cartography*. Revised and enlarged by R. A. Skelton. London: Watts, 1964. 312pp.

Evans, I. M. and Heather Lawrence. *Christopher Saxton: Elizabethan map-maker*. Wakefield: Wakefield Historical Publications; London: The Holland Press, 1979. 186pp.

George, Wilma. *Animals and maps*. London: Secker & Warburg, 1969. 235pp.

Hsu, M-L. 'The Han maps and early Chinese cartography'. *Annals* of the American Association of Geographers, **68**: 45–60 (1978).

Hsu, M-L. 'A Chinese map of the fourth century BC.' Paper presented to the 10th International Conference on the History of Cartography, Dublin, 1983. 5pp.

The Monotype Corporation. 'Precision in mapmaking'. *The Monotype Recorder*, **43**(1): (1964). 52pp.

Nebenzahl, K. (editor). *Atlas of the American Revolution*. Chicago: Rand McNally & Company, 1974. 218pp.

* Numbers in square brackets refer to the bibliographic sections, Parts II and III.

Nordenskiöld, A. E. *Facsimile atlas to the early history of cartography*. Stockholm: 1889. 142pp. Reprinted, New York: Dover Publications, 1973.

North, Gary W. 'Maps: who uses them?' Geography and Map Division, Special Libraries Association, *Bulletin*, **130**: 2–19 (December 1982).

Richeson, A. W. *English land measuring to 1800: instruments and practices*. Cambridge, Massachusetts: Society for the History of Technology, and MIT Press, 1966. 214pp.

Skelton, R. A. Foreword to *Mercator* by A. S. Osley. London: Faber & Faber, 1969. 209pp.

Stevens, Henry N. *Ptolemy's 'Geography'*. Originally published, London: 1908. Reprinted, Amsterdam: Theatrum Orbis Terrarum. N.D. 62pp.

Thompson, F. M. L. *Chartered surveyors; the growth of a profession*. London: Routledge & Kegan Paul, 1968. 400pp.

Wilford, J. N. *The mapmakers*. London: Junction Books, 1981. 414pp.

# 2 Cartographic Information: Its Origins and Utilization

The late R. A. Skelton [319] suggests that four attitudes or approaches to the study of early maps are possible. He regards a map as:

(a) material for reconstructing the physical landscape;
(b) as evidence of human life and organization;
(c) as an illustration of the state of geographical knowledge and thought; and
(d) as a product of cartographic skill and practice.

He goes on to suggest that (a) and (b) are the province primarily of the historical geographer, but also of the archaeologist and the social, economic or political historian; (c) and (d) are the fields of the historian of science and technology. It seems a little surprising that he did not see (c) also as coming within the province of the historical geographer.

How generally do these conclusions tie up with the authorship of the great volume of research that has been done into cartographic development? Antiquaries such as Richard Gough have not only assembled fine collections but bequeathed them to institutional collections; in Gough's case the Bodleian Library, for which he published an inventory and catalogue of his collection. Historical geographers such as J. B. Harley have published a wealth of scholarly research into various aspects of early cartography; the custodians of major map collections such as Edward Lynam, Walter W. Ristow and R. A. Skelton have made outstanding contributions to the literature with research to a large extent based on material in the collections under their care; and last, but far from least, is the contribution made by scholar/collectors. There is a tendency in some quarters to rise up in wrath at the mention of private map collecting but it should be remembered that many of our major institutional collections are largely made up of benefactions from individual collectors. It is interesting to note in this context that J. B. Harley (1972), writing about carto-bibliographies of English county maps, states: 'The enthusiasm of collectors (more than of historians) has resulted in substantial progress in the bibliographical analysis of county maps.' It is surprisingly rare for anyone practising as a professional cartographer to research and publish anything about the history of cartographic development and even in the field of modern mapmaking most of the literature comes from the pens of academic cartographers or geographers.

As historical documents, early maps are of course an important source for present-day research and it is unfortunate that, maps being flimsy in character, many have been lost through general wear and tear. In some cases it is fortunate that excellent facsimiles are available of original maps which have been lost; the outstanding Ebstorf *mappa mundi*, for example, which was destroyed by Allied bombing during the Second World War, is a case in point. Fortunately, too, there is a copious literature on early maps, a large part of it being the product of research by private scholar/collectors. The renowned *Facsimile atlas* [25] of A. E. Nordenskiöld is an outstanding example of a private collection being used as a research tool for the benefit of future students of early cartography.

Serious collecting in England began in the latter stages of the sixteenth century when maps from libraries broken up at the dissolution of the monasteries came into the private collections of antiquaries such as Archbishop Matthew Parker and Sir Robert Cotton. We owe much of our knowledge of important medieval maps such as those of Matthew Paris to their preservation in such collections and their later conservation in the British Library. Another systematic collector of the period to whom we owe a debt for saving so much early material is William Cecil, first Lord Burghley, whose great map collection, described by R. A. Skelton and John Summerson (1971), is still preserved at Hatfield House, home of the Cecils.

For some of our knowledge of the European map trade we can turn to the archives of the great cartographic figures of the day such as the Antwerp typographer, printer and publisher Christopher Plantin. Such records form a valuable documentary source on the Netherlands map industry in the seventeenth century. Abraham Ortelius, also living in Antwerp, amassed a celebrated collection of maps and books that reflected his historical interests, and his correspondence shows how he used agents in different countries to seek out the best maps for him to use in his atlas *Theatrum orbis terrarum*.

Many of the great national libraries of Europe have benefited from the collections of their sovereigns; for example, the British Library preserves two collections of cartographic material—Topographical and Maritime—which were assembled privately by George III.

During the eighteenth century map collecting was a popular pursuit, numerous handbooks being produced for the benefit of collectors. Scholars who collected were generally more interested in the examination of the topographical content of a map and its ability to assist them in reconstructing the past environment than in the map as a specific example of early cartography. In the nineteenth century, however, the study of early cartography took some major steps forward with the publication of two important sets of facsimiles. The Bibliothèque Nationale in Paris had its separate Département des Cartes et Plans administered by Edme François Jomard. Jomard assembled the map collection systematically with a view to 'facilitating comparative study of the ancient manuscript maps scattered throughout Europe'. The items were carefully selected so as to provide clear copies of unique or scarce maps that might deteriorate with time. Skelton (1972) tells us that Jomard's facsimile makes it evident that more names could be read on the world map of Juan de la Cosa than can be deciphered today.

Manuel Francisco de Barros y Souza, the second Viscount of Santarém, was the director of archives in Portugal from 1824 to 1827. He had a passionate interest in early maps and was able to put them to practical use in a political issue. In 1840 sovereignty over part of Senegal was at issue between Portugal and France. Santarém, whose main career was in politics, prepared a memoir supporting the Portuguese case and prepared an atlas of map facsimiles to accompany the memoir. Later Santarém, in pursuit of his studies in cartographic development, enlarged this facsimile atlas from twenty-one to seventy-seven sheets. The maps were printed by lithography, often using colour, but were not faithful facsimiles like those produced today, but hand-drawn copies.

A third scholar, J. G. Kohl from Bremen, pursued research in historical geography and to illustrate this work prepared hand-drawn copies of early maps concerned with the discovery and exploration of America. These he took to the United States in 1854 and with government assistance prepared a series of these copies 'as the foundation of an elaborate catalogue of the early maps of the American continent'. In 1886 the maps were catalogued at Harvard and are now housed in the Library of Congress, where they provide valuable study material for historians and others.

Santarém, Jomard, Kohl—all were pioneers in providing a generous supply of documentary evidence for the student of cartographic development. Nevertheless, their work provided more evidence of content than style for in each case the maps were hand-drawn copies. The distinguished scholar A. E. Nordenskiöld, however, produced two important atlases that faithfully reproduce rare material. Nordenskiöld, geologist and explorer, has been described as the founder of modern historical cartography. An ardent bibliophile, he collected books on a large scale and accumulated one of the finest private libraries covering exploration and the history of map-making. An annotated catalogue of the collection by Ann-Mari Mickwitz and L. Miekkavaara [5] is now available and is of considerable service to students of historical cartography. Nordenskiöld himself, however, made major contributions to research in his *Facsimile-atlas to the early history of cartography* [25] and *Periplus* [235]. In the former he reproduced the most important maps of the fifteenth and sixteenth centuries along with an extensive text, while the latter is 'an essay on the early history of charts and sailing directions'. The major breakthrough provided by these two works, apart from the quality of the scholarship and the wide range of the material reproduced, is that the maps were faithfully printed by photolithography. While some facsimile atlases have themselves become extremely rare, Nordenskiöld's happily are available today in excellent reprints and remain among the most important source works on early cartography.

The finest and most ambitious of cartographic facsimile publications is the sixteen-volume *Monumenta cartographica Africae et Aegypti* (1926–51) [149], which was edited by the Dutch cartophile F. C. Wieder for Prince Youssouf Kamal of Egypt. Wieder specialized in the cartography of Dutch voyages of discovery and of colonial expansion and published numerous papers and books relating to the maps he studied but it is perhaps in his collaboration with the Egyptian Prince Youssouf that he made his most important contribution. The contents of *Monumenta cartographica* are detailed in volume 6 of the *List of geographical atlases in the Library of Congress* [11] and a perusal of these detailed tables of contents immediately shows what a remarkable store of information on early cartography Wieder provided. Another Dutchman, Frederick Muller, an antiquarian book-seller of Amsterdam, for whom Wieder had compiled catalogues in the early stages of his career, was an important influence in stimulating Dutch interest in early maps and atlases. His firm of Fred. Muller & Co. sponsored a series of facsimiles under the title *Remarkable maps of the XVth, XVIth and XVIIth century, reproduced in their original size* (1894). The catalogue/bibliographies of the firm illustrate the extent of Muller's promotion of early maps and atlases; for example, in 1872–75 Muller published his *Catalogue of books, maps, plates on America*, a catalogue that included the largest collection of early atlases ever to be detailed in a bookseller's catalogue and which represented great progress in the bibliographical description of atlases. Muller was inspired in his studies of cartographic development by Bodel Nijenhuis, a collector specializing in the Netherlands, and it is in this trio—Weider, Muller and Nijenhuis—that Koeman [24] says 'we have the personification of the history of the Dutch map collections during the last hundred years'.

Prior to 1916 only the British Museum had made available complete cata-logues of the cartographic material in its care. Philip Lee Phillips had directed the listing of the Library of Congress holdings of maps of America and prepared the first four volumes of the *List of geographical atlases* [11] in the Library of Congress, a list later extended to eight volumes under the editorship of Clara Egli LeGear.

Some other countries issued catalogues of maps and atlases in their libraries and archives and are listed by Skelton [319]. Other specialized catalogues and lists were concerned with specific types of map. Konrad Miller's *Mappaemundi* (Stuttgart, 1895–98), for example, systematically listed and classified medieval world maps.

The last decades of the nineteenth century and the early years of the twentieth constituted a period in which the literature concerned with the history of cartography was steadily expanding, particularly studies related to regional cartography. In this category outstanding contributions were made by Henry Harrisse, whose works [98], [99], [116] remain important source material to this day. In the early twentieth century the student of early mapmaking, and particularly of regional geography or of exploration and discovery, had numerous guidelines available to direct him to suitable sources in pursuit of particular maps. The number of reprints of works mentioned testify to their continuing usefulness today. Furthermore, the literature on mapmaking history covered a number of specific themes (Skelton (1972) lists eight) ranging from the patterns and purposes of medieval world maps, to map projections, the development of conventional symbolism and the application to maps of printing techniques. There could be no question that the study of early maps was firmly established.

## Present-Day Studies of Cartographic Development

Research and study continues to be dominated by historical geographers and by the custodians of map collections—not surprisingly, in view of their easy access to study material. There is also a growing amount of literature aimed at collectors and at a more general market. Authors of such works, such as R. V. Tooley and R. Baynton-Williams, are often involved in the antiquarian map trade.

One of the most ambitious projects currently under way is a massive twelve-volume history of cartography being prepared under the editorship of David Woodward and J. B. Harley. The former, an Englishman who has worked for some years in the United States, has a particularly appropriate background, for unlike many of today's scholars, he gained practical experience of cartography with the Directorate of Overseas Survey, although he later studied at Swansea and at the University of Wisconsin. Woodward's interest in printing processes is reflected in his editorship of *Five centuries of map printing* [261], a collection of essays that is almost unique in concerning itself entirely with the technical aspects of the various map printing processes. Among his long-term projects, Woodward includes a bibliography of maps and city views published in Italy in the sixteenth century. His co-editor, J. B. Harley, is Montefiore Reader in Geography at the University of Exeter and his scholarly contributions to the literature of eighteenth- and nineteenth-century cartography in Britain, of military mapping in the United States, and of the Ordnance Survey are immense. Of particular value, not only to the local historian for whom it was written, but to all who use maps, is Harley's *Maps for the local historian: a guide to the British sources* [198], which reviews six different types of maps and plans as sources of historical evidence. Dr. Harley's companion volume on Ordnance Survey maps and plans—*The historian's guide to Ordnance Survey maps* [788]—examines in some detail the products of the British national survey from the time of its inception. The same body's current output is discussed and illustrated in Harley's authoritative survey *Ordnance Survey maps: a descriptive manual* [787].

No single person has contributed more to the study of early mapmaking than

Professor Dr. Ir. C. Koeman, Professor of Cartography in the State University of Utrecht. He has had a deep and wide-ranging interest in cartography since studying land surveying at the University of Technology in Delft. In recent years he has contributed major bibliographical reference works concerned with Dutch sea and land atlases and with map collections in the Netherlands. His *Collections of maps and atlases in the Netherlands* [24] was designed to contribute to the scientific foundations on which the study of early cartography rests. Koeman aimed to achieve this in the pages of a single volume in three ways: by publishing a list of present-day map collections in the Netherlands; by formulating a schematic outline of the bibliography of atlases published in the Low Countries before 1800; and by studying the character of map collecting throughout the centuries and particularly the history of Dutch map collections. Professor Koeman's wide experience of his subject enabled him to provide a comprehensive picture of the enormous wealth of early cartographic material now preserved in the Netherlands. He later served as an advisory editor in the fine series of facsimile atlases issued by Theatrum Orbis Terrarum (TOT) of Amsterdam and contributed introductory material to several items. His monographs *Joan Blaeu and his grand atlas* [53] and *The sea on paper: the story of the van Keulens and their 'sea-torch'* [54], which originally accompanied facsimile issues in the TOT series, are now available separately. Koeman's major work, however, is the compilation and editing of *Atlantes Neerlandici* [4], a five-volume work published in 1972 by TOT, which is a complete bibliography of terrestrial, maritime and celestial atlases published in the Netherlands up to 1880. It describes 1,017 editions and issues of terrestrial atlases, 680 volumes of maritime atlases, rutters and pilot books, and celestial atlases and charts by sixty publishers. This monumental work, invaluable to all map curators, scholars and collectors, also contains biographies of the most important publishers and mapmakers. In addition to his research and writing on early maps, Professor Koeman lectured on modern cartography at the University of Delft and at the International Training Centre for Aerial Survey.

Among other specialist-contributors to the TOT facsimile series were the late Professor Coolie Verner, Dr. Walter W. Ristow and the late Dr. R. A. Skelton. Verner enjoyed an international reputation, not only for his work on early cartography, but also for his professional activities in the fields of rural sociology and adult education. His contributions to carto-bibliography began around 1950 with studies of the early maps of Virginia—now available on microfilm. He had a special interest in copperplate printing, a topic on which he contributed to Woodward's *Five centuries of map printing* [261]. A selective bibliography of his published work on the history of cartography is available in *Imago Mundi (33*, 100–2 (1981)).

Dr. Walter W. Ristow's contribution to map curatorship and its literature is unrivalled. He served on the staff of the Library of Congress since 1946, becoming its chief in 1968. A Festschrift volume, *The map librarian in the modern world: essays in honour of Walter W. Ristow* [613], includes on pp. 20–46 a list of his publications. These range over many specialist topics in cartography and map librarianship including the very useful *Guide to the history of cartography: an annotated list of references on the history of maps and mapmaking* [3]. This work includes references to a number of carto-bibliographies and other aids to the study of cartographical history. Another outstanding servant of the Geography and Map Division of the Library of Congress was Clara Egli LeGear, who was a member of staff for forty-six years. Her most outstanding published works are volumes 5 to 8 of the *List of geographical atlases in the Library of Congress* [11].

Successive superintendents of the Map Room in the British Museum have made important contributions to the literature of mapmaking, none more so than Edward Lynam and his successor, R. A. Skelton. Lynam's varied interests are summed up in *The mapmaker's art: essays on the history of maps* [260], in which he discussed topics as wide-ranging as 'William Hack and the South Sea buccaneers', 'Saxton's atlas of England and Wales' and the particularly interesting 'Period ornament, writing and symbols on maps'. Dr. Skelton, after retiring from the post of Superintendent of the Map Room in 1967, embarked upon a series of scholarly projects and the publications resulting from his researches are listed in a bibliography on pp. 111–31 of his book *Maps: a historical survey of their study and collecting* [319]. His major work, however, is *County atlases of the British Isles 1579–1850. A bibliography compiled by R. A. Skelton, 1579–1703* [195], which promised to be the definitive volume on British county atlases but sadly was not to be completed by Skelton owing to his untimely death in a motoring accident. The work is designed as an aid to the study of British county atlases, an aim that is achieved in three ways: by establishing the generic relationship between atlases; by tracing the history and successive users of map plates; and by providing sufficient detail for single maps to be assigned to the atlas in which they were originally issued. This last work of Skelton's is an indispensable tool for map curators, librarians, historians of county mapmaking and collectors. (*See* p. 29.)

So far all the contributions to the literature we have considered have come from the pens of academic historians of cartography or from keepers of map collections. A smaller but not insignificant contribution has been made by professional dealers in early maps. P. J. Radford [325] and R. Baynton-Williams [320] have issued popular works for the collector but undoubtedly the major contributions in this sector have been made by R. V. Tooley, a man who combined the virtues of wise dealing, perceptive collecting and enthusiastic scholarship. Numerous volumes in the Map Collectors' Series were compiled by Tooley, including several on the maps of Africa. These were later brought together and published in one volume as *Collectors' guide to maps of the African continent and southern Africa* [147], a book that describes the work of over one hundred mapmakers and lists many examples of their maps relating to Africa. Tooley's major work, *Tooley's dictionary of mapmakers* [43], amply reflects the author's life interest in early maps, for its information 'is based on my observation and study of maps over a period of fifty years'. The number of entries is 21,450 and each provides a minimum of biographical information— sufficient only for identification purposes. Although Harley, in a review in *Imago*

*Mundi* (**32**, 98–9 (1980)), expressed surprise at the 'smallness' of the number of entries, Tooley's work is a remarkable one and a unique reference aid for anyone requiring biographical data about a vast number of mapmakers.

Before we leave sources of information on the history of cartography, mention must be made of the usefulness of map catalogues issued by dealers, a usefulness that is emphasized by the occasional modern reprints of early catalogues, such as *Sayer & Bennett's Catalogue of prints for the year 1775*, a list which includes a considerable section on maps with their contemporary prices [67]. Of the catalogues currently being issued none is better produced than or as informative as those issued by Robert Douwma of London. These beautifully designed volumes include important annotations to each item, often contributed by Tony Campbell.

## Writers on Contemporary Cartography

While the literature on modern mapmaking is possibly not quite so extensive as that on historical matters, there is a formidable amount of information to be gleaned from textbooks, journals, theses, conference reports, yearbooks and the progress and technical reports of government agencies.

Standard textbooks on cartography are a comparatively recent development. Before the 1930s courses had to be taught without the aid of an accompanying text for none was available in the English language. Nowadays there are numerous works at a variety of levels. Generally these are written by academic cartographers or geographers rather than professional mapmakers fully engaged in the commercial or official production of maps and atlases. There is a notable emphasis on small-scale thematic mapmaking, for such an approach is geared to the widest market, that of the student in higher education who is required to prepare maps for a dissertation. In some cases the authors have little or no practical experience of preparing maps themselves. The most notable exception is J. S. Keates, who has produced a definitive work, *Cartographic design and production* [467]. Keates's wide experience of both professional and academic cartography is reflected in a work that deals fully with the graphic and technical bases of cartography, along with map reproduction. In a shorter work, *Practical map production* [468], John Loxton examines briefly the production of topographic maps, looking at generalization and selection, symbolization, the cartographic drawing office, map reproduction and map revision. He concludes with information on map records, index systems, storage, costing and pricing.

The late Erwin Raisz did much to further geographical cartography during the 1940s and 1950s, partly as a result of his authorship of the first recent textbook on the subject: *General cartography* [357]. This volume aimed to help a student to understand the language of maps and to give him a foundation on which he could build if he chose to take up cartography as a professional career.

Among the greatest contributions to the literature and teaching of cartography in recent times are those of Arthur H. Robinson, until his retirement

Professor of Geography at the University of Wisconsin. The first edition of his standard textbook, *Elements of cartography* [359], appeared in 1953. This important work remains the best general approach to the subject and is currently available in a much revised, expanded edition which is now in the hands of two additional authors, Randall Sale and Joel Morrison. It is now completely up to date and includes material on the applications of the computer, automated methods, scribing, aerial photography and so on. In 1952 the University of Wisconsin Press published *The look of maps*, a work that is an expanded version of Professor Robinson's 1947 dissertation subject of 'Foundations in cartographic methodology'. *The nature of maps: essays toward understanding maps and mapping* [360] was written in collaboration with Barbara Bartz Petchenik and is basically a philosophical analysis of the map as a communications system. Robinson's most recent book, *Early thematic mapping in the history of cartography* [300], provides a much-needed narrative history of the origins and development of thematic mapping, a topic previously covered only in scattered papers.

The staff of the University of Wisconsin have done much to widen the scope of cartographic literature. After Robinson and Petchenik comes Phillip C. Muehrcke, an author who brings a lively approach to writing about maps. His *Map use: reading, analysis and interpretation* [598] is written exclusively for the map user, and it examines map use under three major headings: map reading; map analysis; map interpretation. Each is clearly defined and related to the others. So too are the distinctions and interrelationships between the tangible cartographic map and the mental cognitive map. Each chapter is followed by a useful list of suggestions for further reading.

Recently David J. Cuff and Mark T. Mattson, authors who combine academic and professional expertise, have published *Thematic maps: their design and production* [350], a practical work designed to guide the student undertaking a first-year course in cartography in the principles and practices of thematic mapmaking. Like Muehrcke, the authors separate their text under three main headings: choice of symbolization and verbal material; presentation of the map; production and reproduction. They include a section on computer-aided mapmaking and an extensive list of references and suggestions for further study.

In Britain too several volumes on thematic mapping have appeared since the Second World War. These have been designed to aid the geography student in preparing maps and diagrams for a dissertation or the author who is faced with the task of preparing maps without professional assistance. One of the earliest and most comprehensive, though now inevitably a little dated, was Monkhouse and Wilkinson's *Maps and diagrams* [484]. These authors were especially effective in outlining a great variety of techniques of statistical presentation and their book remains an important reference source of methodology. Several books with similar aims followed, their purpose being to aid the amateur rather than the professional mapmaker; Hodgkiss's *Maps for books and theses* [481], Dickinson's *Statistical mapping and the presentation of statistics* [480] and Lawrence's *Cartographic methods* [482] are typical examples of the genre.

## Bibliographical Works

In recent years a small number of authors have attempted the daunting task of providing information about the entire contemporary cartography output: the finished product, rather than the process of producing it. C. B. Muriel Lock, in *Modern maps and atlases* [576], was a pioneer and her mammoth, well-researched work describes several thousand cartographic products and works about cartography. Lock treats her material systematically by continent, country and region, as does Larsgaard [344], whose work largely supersedes Lock within its regional specialization. Unlike these narrative treatments, Kenneth Winch in *International maps and atlases in print* [578] simply provides a systematic listing of maps and atlases in print throughout the world and as such gives the map curator—or anyone involved in the systematic acquisition or study of maps—a unique reference work although now, regrettably, becoming dated.

## Cartographic Journals

Since the close of the Second World War, and particularly since the 1960s, there has not only been rapid development in cartography itself but also a proliferation of books on cartographic matters, and a number of excellent journals have appeared that provide a supplementary source of information about the history of cartography and present-day production and technology. These journals will be discussed at some length in a later chapter.

## The Treatment of Maps and Other Cartographic Material in Libraries

Major map libraries such as the Bibliothèque Nationale, the Library of Congress and the British Library maintain well-established, extensive collections of maps, atlases and other related items. Most universities operate sizeable map collections, normally of current or recent material rather than early maps and atlases, though some major libraries (such as the Bodleian at Oxford) house superb collections of early items. The average university collection is housed in the department of geography and is administered by one or more full-time curators. Local public libraries that serve a sizeable population often have a good collection of standard topographic series and are a prime source for local material. Not all these collections, by any means, are readily accessible to the general public. In England the most important large map collection which is freely open for public use is that of the Royal Geographical Society in Kensington Gore, London SW7 [887]. It is likely, however, that the curators of university collections would be willing to help with specific queries of a serious nature if and when time permits. Their primary duty obviously is to serve the students and staff of their university and it is always sensible and courteous to place a written request when seeking advice or assistance from a university curator.

The RSI section of the Library Association in Britain included a section on maps in *Standards for reference service in public libraries* (1969). Referring to those public libraries which serve a minimum 300,000 population, this section notes that the following provisions should be made: for Great Britain there should be full coverage at $2\frac{1}{2}$ inches to 1 mile, 1 inch to 1 mile, and smaller-scale Ordnance Survey series; full coverage of the local area at 6 inches to 1 mile and 50 inches to 1 mile; 6 inches to 1 mile geology sheets; and complete series of the Soil Survey and Land Utilization series; while for overseas countries there should be the 1:1,000,000 world series, 1:200,000 series for Western Europe, 1 inch to 1 mile for Ireland, and plans of the world's major cities. While these indications are now out of date, and were ever a standard to aim for rather than one that was commonly achieved, they do give some rough idea of the sort of coverage the enquirer might expect to find in a largish British public library.

## References and Further Reading

Harley, J. B. *Maps for the local historian: a guide to the British sources*. London: National Council of Social Service, for the Standing Conference for Local History, 1972. 86pp.

Skelton, R. A. *Maps: a historical survey of their study and collecting*. Chicago and London: University of Chicago Press, 1972. 138pp.

Skelton, R. A. and John Summerson. *A description of maps and architectural drawings in the collection made by William Cecil, First Baron Burghley, now at Hatfield House*. Oxford: Roxburghe Club, 1971. 111 pp.

Tooley, R. V. *Maps and mapmakers*. 4th edn. London: Batsford, 1970. 140pp.

*Note added in proof*

Skelton's major work on county atlases of the British Isles, described on p. 25, is fortunately being continued by Donald Hodson, one of Skelton's collaborators:

Hodson, D. *County atlases of the British Isles*. Vol. 1, *Atlases published 1704–1742 and their subsequent editions*. Tewin, Hertfordshire: The Tewin Press, 1984. 216 pp. 9 plates.

# 3 Who? What? Where?

## National Survey Organizations: British Isles

Cartography in the British Isles remains a small industry employing relatively few people and, as we have noted in Chapter 1, is mainly concentrated in the areas of London, Southampton, Edinburgh, Oxford and Taunton, with tiny pockets elsewhere. The major employers are the national survey establishments, of which Britain has four actively engaged in producing maps from original data. They are the Ordnance Survey (established in 1791 as the Trigonometrical Survey) with its headquarters in Southampton; the Admiralty Hydrographic Department; the Directorate of Military Survey; and the Ordnance Survey of Northern Ireland. On 1 April 1984 the former Directorate of Overseas Surveys merged with the Ordnance Survey and became the Overseas Directorate of the Ordnance Survey.

The past few years have seen a decline in the industry owing to the economic recession. In 1975–76 the total staff employed by the Ordnance Survey amounted to 4,453 but by 1983–84 the number had fallen to 2,814. There has been a marked decrease in the sale of maps at a scale of 1:10,000 and larger, but small-scale sheets remain popular. The famous one inch to one mile series, of which the first sheet of the Old Series was published in 1801, is now a thing of the past. The fine Seventh Series was the last to be published and the scale has now been superseded by the metric 1:50,000. However, the primary purpose of the national survey is to record change in the landscape as it occurs (Harley, 1975) by surveying and providing maps of Great Britain and by keeping existing maps up to date by constant and precise revision. The standard account of the history of the Ordnance Survey is *A history of the Ordnance Survey*, edited by W. A. Seymour, 1980 [792], and a full, authoritative account of the present-day organization, its workings and the whole range of its publications can be found in *Ordnance Survey maps: a descriptive manual* [787]. This comprehensive volume replaces a series of booklets ([789], [790], [791]) that dealt separately with small-, medium- and large-scale maps. The last of these was published in 1957. Harley and Phillips have also described the range of Ordnance Survey publications in a most useful booklet entitled *The historian's guide to Ordnance Survey maps* [788]. From the point of view of acquisitions, the librarian should find much of the information he requires in the Ordnance Survey's general catalogue and he can find out something of the workings, general policy and future plans of the Survey in its series of Annual Reports.

The Overseas Directorate of the Ordnance Survey (as we have seen, formerly the Directorate of Overseas Survey) has been, and continues to be, responsible for much of the mapwork for former British colonial territories and developing countries. The DOS was established in 1946 and many of its maps are among the most stimulating products of modern cartography. Of the six hundred or so maps it publishes annually, 90 per cent were produced from aerial photographs. Between 1980 and 1983 the Directorate produced maps of thirty-nine countries, mainly in the Caribbean, African and Pacific regions. An important function of this body is to provide aid to developing countries in land survey and mapmaking under the U.K. Programme of Technical Co-operation. Senior technical and professional staff are seconded to overseas governments and fifteen to twenty places are available every year for overseas personnel to receive practical technician training in the basic areas of topographic and thematic mapping.

The Admiralty Hydrographic Department with its headquarters at Taunton in Somerset has as its main objective the production of nautical charts that will suit the requirements of the Royal Navy. However the charts are widely used by naval or merchant seamen, fishermen, harbourmasters, pilots and yachtsmen. In addition to the normal range of standard charts, 'special-purpose' charts can be supplied to meet the individual needs of submarine commanders, hovercraft pilots or naval aviators. Supertankers and other large bulk carriers with their limited manoeuvrability are catered for by specially prepared Routeing Charts designed to help in planning the routes of such vessels. Hydrographic detail is collected by special survey ships, each under the command of an officer who is as skilled in survey as he is in seamanship. The collected data are interpreted and processed on board ship before being drawn up and sent to the Hydrographic Office in Taunton for fair drawing and reproduction. Today the Department publishes a range of over 3,000 items with a total of several million sheets. The Hydrographic Department is a member of the International Hydrographic Bureau (established at Monaco in 1921), through which information is freely exchanged, and it is the policy of the Department to make its charts available to the ships of any nation so that they may navigate with greater safety.

The purpose of the Directorate of Military Survey is to serve the needs of the three Armed Services for land maps, aeronautical charts and other geographical data. The Survey has its own map library with holdings of approximately one million maps and charts and is constantly accumulating new accessions that help to provide a bank of geographical data for use in the event of any national emergency as well as for routine internal reference and compilation use. The basic standard scale of military mapping is 1:50,000 but special situations may require the production of supplementary maps outside the usual scale range. For example, the one inch to one mile maps of the Ordnance Survey of Northern Ireland proved to be insufficiently detailed for military operations in the particular terrain and it was necessary to provide a complete coverage of the Province at 1:20,000 and of towns at 1:2,500 (Shelswell, 1982).

The British Ordnance Survey is no longer concerned with maps of Ireland; the Province's own survey, the Ordnance Survey of Northern Ireland, has its headquarters in Belfast while four regional offices maintain a continuous revision system. A new 1:50,000 series is nearing completion, as is the Survey's objective of publishing large- and medium-scale mapping on the Irish Grid system.

## Other official and governmental agencies in the United Kingdom

Several other agencies in the United Kingdom are responsible for the type of map production that involves original research rather than simply producing maps derived from Ordnance Survey data. Generally they are there to serve specific users and their output is mainly thematic. The Department of the Environment produced a series of planning maps covering a wide range of geographical themes. Each map was in two sheets—North and South Britain—and the series as a whole formed the nearest approach Britain has had to a national atlas. Recently much of the Department's map production has been geared to internal use, with some exceptions such as the provision of detailed maps showing river quality, effluent disposal and biological data for the water industry.

The British Geological Survey publishes 1:50,000 geological maps which replace the one inch to one mile series as well as a series at 1:250,000 to cover the United Kingdom and its continental shelf. Some 130 maps per year are produced at the field survey scale of 1:10,000 and these are available as film positives from which dieline copies can be made as required. Additionally a number of special sheets at the 1:25,000 scale are available and maps accompanying reports have been prepared at this scale for the Department of the Environment to assess resources of sand and gravel, conglomerate and limestone.

The Macaulay Institute for Soil Research is concerned with the systematic soil survey of Scotland and over thirty soil survey maps have already been produced as well as twenty-three land-use capability maps, all at 1:63,360. Additionally, national soils and land use capability maps in five sheets at 1:250,000, a number of dieline soil maps for limited circulation, and several derivative small-scale maps have been published.

The Soil Survey of England and Wales was established in 1939 and since 1946 has had its headquarters at the Rothamsted Experimental Station, Harpenden. The main products have been a one inch to one mile series, which will eventually be replaced by 1:50,000 sheets; a 1:25,000 series accompanied by land-use capability maps; and a 1:250,000 soil map in six sheets, which forms the first systematic fully comprehensive inventory of the soil resources of England and Wales.

Maps based on data culled from the National Censuses are prepared by the Office of Population Censuses and Surveys: Census Division. Thematic maps show the geographical detail contained in 130,000 census enumeration districts and are published in census reports. Wall charts are also produced for free distribution to educational establishments.

Other agencies producing maps involving some form of original survey include the oil company BP; the Institute of Hydrology, whose main fields of interest are currently flood studies and river maps, the latter coloured according to pollution and with the width of the river made proportional to the flow; and the National Remote Sensing Centre.

## British agencies producing largely derived maps

Generally speaking it is commercial organizations that produce derived general-purpose maps and atlases though a list of eighty-seven publishers of maps in the United Kingdom (*Cartographic Journal*, **21**, 47 (June 1984)) includes bodies such as the Forestry Commission, the Scottish Tourist Board and the City of Westminster. While many of the strictly commercial firms are very small, others, such as John Bartholomew & Son Ltd. of Edinburgh, have a long record of fine and varied map production and maintain relatively large establishments. One of the best-known firms, the George Philip Group, publishes an impressive list of maps and atlases at various levels including special-purpose atlases such as a *Primary atlas for Tanzania schools* in Swahili; a *Certificate atlas for the Caribbean*; and a *Home reference atlas for Canadians*. Another name familiar in Great Britain is that of Wm. Collins Sons & Co. Ltd., a firm which publishes numerous road atlases as well as the extensive range of Collins–Longman School Atlases, again including special atlases designed for use in Turkish, Nigerian and Zimbabwe schools. The Geographers' A–Z Map Co. Ltd. specializes in the very useful series of A to Z street plans, usually readily available in a good local bookshop. Medium-sized establishments, of which Clyde Surveys is a good example, undertake a wide range of survey and map production tasks from road construction mapping programmes to the production of satellite maps. Smaller firms such as Geoprojects of Henley on Thames offer a variety of work from the preparation of medium- and small-scale maps to atlases, road maps and town plans.

Information about the products of both official and commercial mapmaking agencies is easily obtained by perusing a copy of their general catalogue, while information concerning the activities and future schedules of official bodies can often be found in their Annual Reports or in the major journals. A particularly interesting series of reports on individual establishments is currently being featured in the issues of the British Cartographic Society's biannual *Newsletter* [406].

A good deal of stimulating cartography is carried out by free-lance practitioners, often for advertising purposes, and the setting up of small cartographic units, manned by anything from one to six staff, in higher educational establishments has been an interesting post-war development. The emphasis in this sector is firmly on thematic and statistical mapmaking, the maps normally being prepared to illustrate academic research and often published in scientific books and journals. Maps are also produced to serve as visual aids, either in the form of large display maps or for reproduction as 35 mm colour transparencies.

## Cartography in Western Europe

Britons seeking a map of one of the European countries often rely on the commercially produced series published by famous names such as Hallwag, Michelin, Firestone, Kummerley and Frey or Freytag-Berndt. Prospective customers are perhaps less aware of the enormous number of maps issued by official cartographic establishments, mainly because such maps are less readily available via retail outlets. Most European nations have major map series at 1:25,000, 1:50,000 and 1:100,000 but the standards of content, style and production vary widely. Generally speaking the cartographical organization within each nation follows the British pattern, with the national body making the primary surveys and a number of commercial companies using such data under licence to prepare derived maps.

In the Federal Republic of Germany, however, the pattern of centralized survey is not followed, for as mentioned in Chapter 1 cartography is generally the responsibility of land-survey administrations within each of the *Länder*. Such a system inevitably means that there are problems of standardization, and to iron out discrepancies between the mapping bodies a working group composed of members from Federal bodies and from each *Land* was set up. This group was also concerned with the general problems of preparing small-scale maps, for the various Land Survey bodies were too fully occupied with post-war large-scale surveys for redevelopment to be able to take on small-scale work also. National series have now been published at 1:200,000, 1:500,000 and 1:1,000,000.

The fine traditions of Italian mapmaking are maintained by the Istituto Geografico Militare, established in Florence in 1872, and the Swiss National Survey, Eidgenössische Landestopographie, founded in 1838, produces superb map series of a country whose terrain lends itself to the making of visually attractive maps. In sharp contrast, the flat Netherlands landscape seems less likely to inspire the making of beautiful maps but the Netherlands remains one of the world's leading cartographic nations with outstanding achievements in the academic teaching of cartography, in the design of cartographical instruments and equipment, and in the splendid map series of the national agency, Topogra-fische Dienst of Delft.

The varied terrain of Scandinavia has produced some stimulating mapmaking from the official bodies of Denmark, Norway and Sweden. The official Danish agency, the Geodaetisk Institut, was established in 1928 as a result of the combining of two earlier bodies, Den Danske Gradmaaling (Royal Danish Arc Survey) and Videnskabernes Selskab (Royal Danish Society of Science and Letters). Current Danish series range from a nationwide 1:20,000, which covers the whole country in 835 sheets, to regional maps at 1:300,000 and 1:500,000. Series at 1:25,000, 1:50,000 and a 1:200,000 motoring set are also available. Norway's official agency, Norges Geografiske Oppmåling, publishes topographi-cal series at 1:25,000, 1:50,000 and 1:100,000 in addition to some superb tourist sheets such as that of the Hardangervidda at 1:200,000. Sweden differs from most countries in having three major cartographic establishments—Generalsta-bens Litografiska Anstalt (GLA), the Esselte Map Service and Rikets Allmänna Kartverk—and is well to the fore, not only for the making of topographic series, but also for statistical mapping, the economic and population maps of W. William-Olsson being renowned examples of their kind.

Greece also has three major cartographic agencies, with the main series of national maps including topographic series at 1:20,000 and 1:50,000. Portugal, a nation with a great tradition in cartography, has national series at standard scales issued by the Instituto Geográfico e Cadastral and Spain, another country with three organizations responsible for national mapmaking, has a basic series at 1:50,000 which covers the country in more than one thousand sheets.

France's national map series are published by the Institut Géographique National (IGN) and as befits a nation that had a fine map series covering the whole country as early as the eighteenth century, there are today a number of fine series, the basic being the 1:20,000.

The cartographic organization of other countries in Western Europe tends to follow a similar pattern and the addresses of the official mapmaking bodies will be found in Part IV. These agencies will supply catalogues showing their series of national maps and special sheets but are unlikely to be willing to provide a great deal of information about their history, organization or prospects. There are, however, other sources to which one can turn. *Modern maps and atlases* [576] by C. B. Muriel Lock has already been referred to as a mine of information on the cartographic output of all the nations of the world and includes excellent lists of references for further reading. The *International Yearbook of Cartography* [389] occasionally features articles on specific mapmaking agencies both official and commercial, as do the various cartographic journals.

## Commercial mapmaking in Western Europe

Like Britain, the other countries of Western Europe have flourishing commercial mapmaking industries with over forty private firms existing in the Federal Republic of Germany alone. The number of automobile maps produced in Germany is well over ten million and with such an output it is important that road maps should not only be highly functional but should be well designed. The products of Mair's Geographical Publishing House of Stuttgart have been especially influential in this sector with two important issues: first the *Deutsche Generalkarte* 1:200,000, a map series of uniform size and of uniform standard of execution; secondly, *The Grosse Shell-Atlas of Germany and Europe*, a work whose stated aim is 'supplying for any given need, the proper map at the most convenient scale'. A paper in German by Volkmar Mair (1963) appeared in the *International Yearbook of Cartography* and provides detailed information about the Mair automobile maps. The same, 1963, issue of the *International Yearbook* contained an article whose first section describes Michelin road maps and whose second the series of red and green Michelin guides. The cartographic products of the Michelin Company are now familiar everywhere and the paper relates how these products have been developed since 1900, when André Michelin wrote his first car travel guide book and followed it up thirteen years later with the first Michelin road map for travellers by car.

The activities of the Istituto Geografico de Agostini-Novara are described by U. Bonapace in the same issue of the *Yearbook*. This organization has, over the past eighty years, produced a varied range of cartographic material including road maps of Italy, atlases, thematic maps, relief maps and globes. The article shows how the Institute has gone through a great variety of technical experiences over the years, passing through hand drafting on stones to the photographic setting of names, and indicates the problems of training cartographic staff to meet continuing new challenges.

The private sector in Europe is so extensive that it is possible to mention only a very few establishments. Among the most interesting is the Bollman Bildkarten Verlag of Braunschweig, a firm founded by Hermann Bollmann and specializing in the most beautiful, detailed perspective plans of historic towns and cities. German firms such as the Kartographisches Institut Bertelsmann and the Westermann-Verlag are renowned for educational atlases; Falk-Verlag of Hamburg specialize in town plans; the Austrian firm of Freytag-Berndt have a varied output of walking maps, regional maps, town plans and automobile maps; R. De Rouck of Brussels produce splendid town plans and route maps. These are only a small selection of firms in the commercial sector of European mapmaking. Details of their products will be found in their individual catalogues and also in the comprehensive catalogue published by Geo Center of Stuttgart, a monumental reference tool that is regularly updated [629].

Worthy of mention also are the renowned panoramic maps of Alpine centres painted by the Innsbruck artist, Professor Heinrich Berann. While these generally appear in the tourist literature of the various Alpine resorts, a good selection is available in Berann's book, *Die Alpen im Panorama*, written with H. A. Graef and published in 1966 by Verlag Weidlich in Frankfurt [737].

# Eastern Europe and the U.S.S.R.

The cartographic achievements of countries of Eastern Europe and of the U.S.S.R. are difficult to discuss owing to the difficulties for Westerners of obtaining full details of scales and map series, and of course of procuring sufficient examples. Apart from 'tourist' maps little cartographic material is available to the West. Lock, however, discusses Soviet mapmaking and atlas production on pp. 258–65 of *Modern maps and atlases* [576]. Probably more is known of Hungarian cartographic activities than those of other countries within the Soviet bloc. International map exhibitions are held regularly in Budapest and Hungarian cartographers collaborate with other nations in mapmaking projects such as the *Atlas international Larousse*. The Kultura Hungarian Trading Company for Books and Newspapers exports road maps, guidebooks and atlases. The State Institute for Geodesy and Cartography in 1965 inaugurated a topical cartographical information service entitled *Cartactual* [378] under the editorship of Dr. Sandor Rado. The bimonthly issues of this publication consist of looseleaf sheets of maps showing topical items of geographical information throughout the world such as boundary changes, changes in placenames, changes or expansion to transport systems, new industrial plants and so on. The two-colour map sheets provide one of the most important sources of topical geographical information, not only for cartographers, but also for librarians and others.

# Former Colonial Territories

The influence of earlier colonial administrations is strongly imposed on many developing countries within the Third World; British cartographic traditions are maintained throughout the Commonwealth and the basic surveys of the French are being developed in former French territories in Africa.

Within the Commonwealth countries there is a wealth of varied terrain, climate and land use which presents a resounding challenge to mapmakers. The Directorate of Overseas Survey has been responsible for a great deal of valuable survey in former British possessions, while some territories have their own fine cartographic traditions. India's, for example, date back as far as 1767 when Major James Rennell was appointed Surveyor General of Bengal. The early days of survey in India are discussed by Heaney (1968). The Survey of India now has headquarters in Dehra Dun and is responsible for the land surveying of the subcontinent while the Naval Hydrographic Branch carries out hydrographic surveys. Standard series of topographic maps are at 1:50,000 and 1:250,000 though other scales are also produced. A full catalogue of publications is available.

The British tradition can also be seen in the fine cartography of Egypt, where the Survey of Egypt has topographical map series at 1:25,000, 1:100,000, 1:250,000 and 1:500,000. The unusual character of the Nile Valley called for an individual approach to cartographic presentation. Contouring, for example, had to be extremely close, that of the 1:25,000 and 1:100,000 series being at one-metre intervals. The narrow strip of cultivated land along the banks of the Nile was presented in three shades of green: pale green for those areas independent of irrigation; a darker green for irrigated, cultivated areas; and very dark green for wooded areas, which were overprinted with symbols indicating types of vegetation. The *Atlas of Egypt*, prepared by the Survey of Egypt in 1928, was one of the best and most comprehensively thematic atlases of its day.

The Directorate of Overseas Survey carried out basic topographical mapping programmes in former British territories such as Sierra Leone, Zambia, Gambia, Uganda, Kenya and what was then Tanganyika. These programmes have been generally maintained by post-colonial survey agencies such as the Department of Lands and Survey in Entebbe and the Survey of Kenya in Nairobi.

For the Republic of South Africa the official mapping body is the Trigonometrical Survey, which produces a basic topographic series at 1:50,000 and other series at varying scales. Information about the state of cartographic production in the country is available in the *Catalogue of maps* issued by the Government Printer in Pretoria.

The Institut Géographique National carried out systematic surveys in Morocco, Algeria, Mauretania, Mali and other former French colonial territories in Africa, and early Belgian influence is seen in maps of the economically important Katanga region of Zaire, which was formerly mapped by the Belgian Military Geographical Institute.

The organization of cartography in Australia differs from the normal pattern of centralized control in that Federal Survey is in the hands of the Director of National Mapping, Department of National Development, while each state has its own mapping authority which prepares large-scale administrative and cadastral maps along with general-purpose maps and tourist material. Australia and New Zealand are well served, as are the Americas, by Larsgaard's *Topographic mapping of the Americas, Australia and New Zealand* [344], in which Part I is an introductory essay on topographic mapping, Parts II, III and IV are country-by-country surveys of topographic mapping past and present, and Part V a bibliography of more than 800 citations. For New Zealand the official mapping organization is the Department of Lands and Survey, which is responsible for all land survey and has a basic topographic series at one inch to one mile (to be replaced by a metric 1:50,000). Philip L. Barton, Map Librarian at the Alexander Turnbull Library, Wellington, has compiled 'A bibliography of material relating to the surveying and mapping of New Zealand' (1977).

Canada's map series are produced under the auspices of the Department of Energy, Mines and Resources. The National Topographic System includes series at 1:25,000, 1:50,000 and 1:250,000. A full analysis of Canadian maps, many of which are superb examples of cartography at its best, is contained in *The maps of Canada: a guide to official Canadian maps, charts, atlases and gazetteers* by N. L. Nicholson and L. M. Sebert [345].

# The United States of America

The topographical series of the United States are in the hands of the U.S. Geological Survey, a body which, in 1879, amalgamated four earlier agencies that had been concerned in both topographical and geological mapmaking. If it seems rather unusual that topographic maps should be the responsibility of the Geological Survey, the reason why they came to be so is interesting. As a result of the important mineral deposits in the west, geologists demanded accurate topographical maps that would form a reliable base on which their findings could be superimposed. A Topographic Branch of the Geological Survey was formed in order to satisfy this requirement. The Geological Survey, which is one of the most generous in providing information about its activities, issues a booklet simply entitled *U.S. Geological Survey*, which describes the resources, functions and services that are its responsibility. Within the Survey there are Administrative and Publications Divisions and four operating divisions: Topographic; Geologic; Water Resources; and Conservation. Index maps of the map series for each state are available free of charge on application to the U.S. Geological Survey, Washington, D.C. 20242. A Map Information Office is maintained in the General Services Administration Building, Room 1028, F Street between 18th and 19th Streets, N.W., Washington and is open to the public for queries on the sources and state of mapping, on geodetic control data, and on aerial photographs. A reference library of four hundred thousand volumes is also open to the public and is housed in the same GSA Building. A collection of over one hundred thousand photographs of geological and cartographic interest is maintained and prints of these photographs may be obtained, for a charge, by writing to the U.S. Geological Survey, Federal Center, Denver, Colorado 80225.

Apart from the official surveys there is a particularly thriving commercial mapmaking sector in the United States which produces a wide variety of products. Around 120 companies are involved and Jon M. Leverenz (1974) separates them into three categories based on the complexity of their production techniques and the sophistication and variety of their products. Around seventy firms produce one- or two-colour street maps or tourist maps of local interest that require little interpretation or generalization from whatever base material is used. Little creative work is involved and the product is designed to meet the needs of people in small urban settlements and rural areas. The second category produce similar products but the map preparation involves greater modification and more sophisticated modification of the source materials. In this case the map users tend to be banks, real-estate companies and publishers in larger urban centres. The third category's extensive range of production includes globes and wall maps, road maps, and general and thematic atlases. The full spectrum of operations involving research, compilation and drafting, and full-colour printing is involved. The market for this category is international as well as national and although there are only about fifteen firms involved they employ almost half the total number of cartographers in the commercial sector.

Among the most familiar American maps throughout the world are the periodic map supplements that appear with the *National Geographic* magazine. As well as the very distinctive reference maps, crammed with place names and verbal annotations, the NGS issues maps on historical themes and on specialized topics such as the beautiful maps of the ocean floors painted by the Alpine panoramic artist, Heinrich Berann.

## Central and South America

The state of cartography of Mexico, Central America, the West Indies and South America is discussed in *A catalogue of Latin American flat maps, 1926–1964* (University of Texas Institute of Latin American Studies, Guides and Bibliographical Series, No. 2), a work that was designed as a follow-up to a *Catalogue of maps of Hispanic America* (American Geographical Society, 1930–32). A useful feature is the inclusion of appendices that list the names and addresses of establishments from which maps and information concerning maps may be obtained. Freeman (1963) also summarizes Latin American cartography, an interesting new development of which is the Pan American Institute of Geography and History's Hemisphere Mapping Project, which aims to complete uniform coverage of the continent at 1:250,000 by the end of the 1980s. The most recent survey of topographic mapping is that of Larsgaard [344].

Mexico's official survey and mapping agency is the Dirección General de Geografía y Meteorología, which has issued a number of thematic maps of the whole country as well as topographical map series, the main series being at the 1:100,000 scale. British Honduras (Belize) was originally mapped by the Directorate of Overseas Survey at 1:50,000, which is the basic scale for Central American countries in general. Information about the state of mapping in Costa Rica, whose official agency is the Instituto Geográfico de Costa Rica, may be found in Albert E. Palmerlee's *Maps of Costa Rica: an annotated bibliography* (Lawrence: University of Kansas Libraries, 1965), a work whose American origin perhaps reflects the influence on Latin American mapmaking of the Inter-American Geodetic Survey, a branch of the U.S. Army Map Service.

Former colonial influence can be seen in parts of South America. For example, the topographic series of British Guiana (now Guyana) and French Guiana were maintained until independence by the Director of Overseas Survey and the Institut Géographique National respectively. Cartographic production in the vast territories of Brazil is detailed in the *Bibliografia cartográfica do Brasil* and for the northeastern region in *Bibliografia cartográfica do Nordeste* (Superintendencia do Desenvolvimento do Nordeste, Divisão de Cartografia, 1965). The cartography of the Amazon region is listed in Isa Adonias's *La cartografia da região amazônica: catálogo descritivo; 1500–1961* [123].

## The Far East

In China all survey, production and publishing of maps is carried out by the

National Bureau of Surveying and Mapping, whose annual sales exceed thirty million copies. The basic national topographic coverage at 1:50,000 (1:100,000 in the remoter areas) and derived smaller-scale mapping is largely complete, although not very widely known in the West.

Japan, as might be expected, also has maps of excellent quality. The basic topographic scale of the Geographical Survey Institute, Ministry of Construction, is 1:50,000 and other series are available at 1:10,000, 1:25,000 and a small-scale series at 1:500,000. For the Hokkaido region land utilization maps have been issued at 1:200,000. Information about mapmaking in Japan may be found in the *Journal of Geodetic Survey of Japan* and in the *Journal* of the Japan Cartographic Association.

The Geographical Survey Institute was set up in 1958 in the Republic of Korea to provide map series at 1:50,000 and 1:100,000 for use in government planning and at 1:200,000 and 1:500,000 for public use. Information about Korean cartographic production is contained in the various issues of the national bibliography.

For Indonesia the Geographical Institute, Directorate of Topography, Ministry of Defence in Djakarta is responsible for producing maps for military purposes as well as general topographic series. Bali has a topographic series at 1:25,000, Sumatra a 1:50,000 series and Java also has 1:50,000 coverage.

Information about the cartography of the Philippines is to be found in *List of maps, charts and views of the Philippine Islands in the Library of Congress* (Washington, D.C.: Government Printing Office, 1903) and in *Philippine Cartography (1320–1899)* by Carlos Quirino [134].

The Institut Géographique National had carried out cartographic operations in Cambodia, Laos and Vietnam until the respective governments assumed responsibility for the mapping of their territories in 1955. The Directorate of Overseas Survey prepared maps at 1:25,000 and 1:50,000 for parts of Malaysia, and the Federation of Malaya Survey Department has produced a series at 1:63,360 in addition to land utilization maps at 1:25,000.

The foregoing is a brief survey of the world's mapmaking establishments; only a large volume such as Lock's *Modern maps and atlases* [576] could attempt any kind of thorough coverage. The same is true for the establishments that are responsible for housing and maintaining collections of maps for consultation.

## Map Collections

Maps are available for consultation in a number of different types of establishment. The largest and most important collections are generally associated with national libraries or with great universities and within such collections may be found treasures from every age of mapmaking: unique manuscript maps, early printed atlases, globes and occasionally scientific instruments used in the process of making maps. It is important to stress that such collections are not generally freely open for public consultation, the Map Room of the Royal Geographical Society in London being one of the few to which the public has ready access. It is sensible and courteous always to submit a letter of enquiry to the Curator of Maps or Superintendent of the Map Room setting out one's problems and requirements: area of interest, purpose for which the map is required, scale and so on.

Many universities maintain map collections of varying range and size, usually housed in the department of geography and under the care of a professional map curator, although specialized collections of maps may be found in the department of geology, planning, oceanography or elsewhere. Valuable early atlases and maps are more likely to be found within the university library, sometimes housed within special collections and occasionally, if they form part of a particular bequest, scattered among other volumes. Many large collections have profited from benefactions by specialist collectors of early material. The A. E. Nordenskiöld collection in the University of Helsinki, for example, is a rich source of historical material for the student of cartographic development; the German historian Christoph Daniel Ebeling created the first important American map collection, which is now housed in the Library of Harvard University; the important map collection of Richard Gough is preserved in the Bodleian Library, Oxford and that of the cartographer d'Anville in the Bibliothèque Nationale. In the United States the Library of Congress map collection has had the benefit of bequests from numerous collectors such as Lessing J. Rosenwald and Melville Eastham.

For the best use to be made of such a wealth of material it is essential that it should be cared for by specially trained staff. It is surprising, therefore, to learn that prior to 1950 map curators 'had to be self-educated, and only too often . . . the needs of map collections were low in the list of priorities as conceived by library administrators' (Helen Wallis in 'Map librarianship comes of age' from *The map librarian in the modern world* [613]). Dr. Wallis goes on to say that the impetus to putting map curatorship on a firm professional footing was the enormous increase in map production during the Second World War and the large consignments of cartographic material that were passed on to map libraries at the close of hostilities. The United States was to the fore in establishing a map curators' group within a national library association, the Geography and Map Group of the Special Libraries Association being set up in 1941. This group's *Bulletin* [375] is now a sizeable quarterly publication and one of the most important sources of information about map curatorship and cartography in general. Similar groups have now been established elsewhere, such as the Australian Map Circle, founded in 1973, and the New Zealand Mapkeepers' Circle, set up a little later in 1977.

Articles describing the history, resources and services of individual map collections within higher-educational establishments are often found in the *Bulletin* of the Geography and Map Division, Special Libraries Association. For example, in No. 107 (March 1977) David C. McQuillan wrote on the 'History of the map collection at the University of South Carolina' and in No. 102 (December 1975) Frances K. Drew, Map Cataloger at the Georgia Tech

Library, Atlanta, related the twenty-five years of growth and development of the map collection under her care.

Libraries in the major cities and towns usually maintain some form of map collection, either of general cartographical material or, perhaps more frequently, of local interest only. Some collections within this sector are of major importance; one of the most unusual is the Newberry Library in Chicago which established a Center for the History of Cartography in 1972 as a result of an endowment from Mr. Hermon Dunlap Smith, a member of the Newberry Board of Trustees and a keen map collector. As well as housing an extensive and varied map collection the Center encourages scholarship in the history of cartography field and offers courses for map collectors and librarians, mounts the regular series of Nebenzahl Lectures and publishes significant works such as *Five centuries of map printing* [261]. The Free Library of Philadelphia also houses an important collection of maps as well as works on the principles of cartography, gazetteers, placename volumes, guidebooks and geographical works. The librarian formerly in charge of the map collection, J. B. Post, also made it his responsibility to build up a comprehensive collection of books on map librarianship, including works on cataloguing and classification as well as on the automation of map collections.

County and local record offices represent a further important source of cartographical information, usually preserving a representative collection of local material varying from early estate maps to modern large-scale urban plans. Outstanding among record offices in Britain is the Essex Record Office in Chelmsford, whose *Catalogue of maps in the Essex Record Office, 1566–1860* by F. G. Emmison (Chelmsford: Essex County Council, 1947) set standards that have not been attained elsewhere. Essex has provided a further service by issuing excellent facsimile reproductions of some of the more important maps from the collection.

Many fine collections of early maps remain in private hands though some, such as the George Allen Collection in the Lancashire Record Office at Preston, are looked after by professional custodians. One of the most important collections in Britain is that assembled by William Cecil and housed at Hatfield. It is described by R. A. Skelton and John Summerson in *A description of maps and architectural drawings in the collection made by William Cecil, First Baron Burghley, now at Hatfield House* (Oxford: Roxburghe Club, 1971).

One of the greatest contributions to the study of the history of cartography was made by Dr. Ir. C. Koeman, whose *Collections of maps and atlases in the Netherlands: their history and present state* [179] aimed to 'contribute to the improvement of the scientific foundations on which the study of early cartography is built . . .' in three ways: first, by publishing a list of present-day map collections in the Netherlands; secondly, by formulating a schematic outline of the bibliography of atlases published in the Low Countries before 1800; and thirdly, by studying the character of map collecting throughout the centuries in general and by describing the history of Dutch map collections in particular. As the range of material in the Netherlands is large Koeman's documentation of this rich cultural heritage is invaluable for anyone wishing to look into the cartographic resources of the country as well as for any serious student of cartographic history in general.

Elsewhere the literature on map collections is confined to scattered papers in cartographic and librarianship journals with the exception of an increasing number of directories of collections. In volume 17, No. 2 of the *SUC Bulletin* [377] the editor, writing in a section labelled 'Curators' Corner', lists eight such directories. In addition to a *World directory of map collections* by Walter W. Ristow [664], directories are listed for Australia, Canada, France, Germany, the Netherlands, the United Kingdom and the United States. For the United Kingdom Barbara A. Bond assembled *A directory of U.K. map collections* [670], a twenty-eight page booklet published in London by the British Cartographic Society in 1983.

## Facsimile Maps

Not all early maps, and particularly unique manuscript maps, are easily accessible, and good-quality facsimile reproductions play a useful role in the study of early cartography and in its teaching. Broadly speaking, facsimile maps may be divided into two categories: first, those produced for historians, geographers, students of the history of cartography, discovery and exploration; and secondly, those produced for a public interested in them mainly as decorative objects. Many of those issued in the first category are accompanied by informative text, outlining particular attributes of the map, assessing its importance in cartographic development, providing biographical details of the mapmaker, and offering guidance for further reading. Hodgkiss (1978) has outlined the history of facsimile production, discussed the range of facsimiles available today and listed the available facsimiles of important maps and atlases.

## References

Barton, Philip L. 'A bibliography of material relating to the surveying and mapping of New Zealand'. Special Libraries Association, Geography and Map Division, *Bulletin*, **109**: 24–32 (September 1977).

Bonapace, Umberto. 'La production cartographique de l'Institut Geographique de Agostini, buts et problèmes actuels'. *International Yearbook of Cartography*, 1963, pp. 157–162.

*Cartographic Journal*. 'Cartographic activities in the United Kingdom, 1980–83'. *The Cartographic Journal*, **21**: 33–54 (June 1984).

Freeman, P. H. 'An inventory of Latin-American mapping'. Special Libraries Association, Geography and Map Division, *Bulletin*, September 1963, pp. 3–5.

Harley, J. B. *Maps for the local historian: a guide to the British sources*. London: National Council of Social Service, for the Standing Conference for Local History. 1972. 80pp.

Harley, J. B. *Ordnance Survey maps: a descriptive manual*. Southampton: Ordnance Survey, 1975. 200pp.

Heaney, G. F. 'Rennell and the surveyors of India'. *The Geographical Journal*, **134**(3): 318–327 (September 1968).

Hodgkiss, A. G. 'Facsimiles of early maps'. *SUC Bulletin*, **12**(2): 1–12 (1978).

Leverenz, Jon M. 'The private cartographic industry in the United States: its staff and educational requirements'. *The American Cartographer*, **1**(2): 117–123 (1974).

Mair, Volkmar. 'Strassenkarten aus Mairs Geographischem Verlag'. *International Yearbook of Cartography*, 1963, pp. 163–79.

Michelin. 'Les services de tourisme du pneu Michelin'. *International Yearbook of Cartography*, 1963, pp. 171–180.

Shelswell, M. A. 'Eyes right—map production in a military survey unit'. *The Cartographic Journal*, **19**(1): 34–45 (June 1982).

Skelton, R. A. *Maps: a historical survey of their study and collecting*. Chicago and London: Chicago University Press, 1972. 138pp.

Skelton, R. A. and John Summerson. *A description of maps and architectural drawings in the collection made by William Cecil, First Baron Burghley, now at Hatfield House*. Oxford: Roxburghe Club, 1971. 111 pp.

Tooley, R. V. *Maps and mapmakers*. 4th edn. London: Batsford, 1970. 140pp.

# 4   The Literature of Cartography

Historians of cartography are admirably served by an extraordinary wealth of scholarly literature dealing with the development of maps and mapmaking from earliest times. Students of contemporary cartography have not been quite so well catered for, particularly as far as the technical side of mapmaking is concerned, though there has been a dramatic increase in the literature of cartography since the end of the Second World War. The tendency has been to divide the broad field of cartography into two major categories: topographic on the one hand, and thematic or 'special-purpose' on the other. Much of the technical literature, certainly as far as books are concerned, has displayed a bias towards thematic cartography and to the theories and principles rather than to the practical applications. That it has done so is possibly a reflection of the fact that many of the authors of books on cartography have been academic geographers rather than professional cartographers. There are of course a number of works that attempt to encompass the whole spectrum of cartography, examples being the excellent *Elements of cartography* by Robinson, Sale and Morrison [359] or *General cartography* by Raisz [357]. While the former deals exceptionally well with both the technical and theoretical aspects of mapmaking, authors such as Raisz or Monkhouse and Wilkinson [484] are more concerned with the graphical presentation of statistics in map form—not surprisingly, as their books are primarily designed to serve the needs of undergraduate geography students in higher educational establishments. One of the few to delve deeper into the technology of mapmaking is J. S. Keates's *Cartographic design and production* [467], a work that is unrivalled in its analysis of the main aspects of cartography: metrical, graphical and technical. Keates wrote primarily for those who intend to make a special study of cartography with a view to taking it up as a career and secondly for those in disciplines that have some special relationship with cartography: surveyors, civil engineers, geographers, for example. Unlike most of the other works on modern mapmaking it is not designed for the amateur mapmaker.

Aside from works in book form, however, there is a wide range of literature covering both the theoretical and technological aspects of mapmaking in the form of scattered papers in the professional journals of cartography and its related disciplines. Some of these papers will be cited in Part III, particularly where no work in book form is available on some specialized aspect of mapmaking.

An important source of information on technology is to be found in the technical reports issued by official, commercial and academic mapmaking bodies. Selective lists of such reports are to be found in issues of the *Bulletin* [375] of the Special Libraries Association, Geography and Map Division. These lists provide details of current geography and mapping reports which are obtainable either in paper or microfiche format from the United States Department of Commerce, National Technical Information Center, 5285 Port Royal Road, Springfield, Virginia 22161. The range of topics is wide: if a random sample is taken from the list appearing in *Bulletin* 125 for September 1981, the subjects found there include automation and data banks; terrain analysis; fish mapping; underwater mapping; vegetation mapping; snow mapping using space imagery, and the mapping of earth fissures in Nevada. Technical reports are also detailed in lists issued by the particular bodies concerned; for example, *New publications of the Geological Survey* lists not only the topographic and specialized maps published by the U.S. Geological Survey during a particular month but also the professional papers, bulletins, Geographic Names Information System products, water resources investigations, and open file reports. The annual reports of official bodies such as the Ordnance Survey are a useful means of keeping up to date with current developments in production and technology concerning the particular issuing body. The first Ordnance Survey Annual Report was issued in 1855–56 and further reports were submitted to Parliament until 1921–22 when the report ceased to be a Command Paper, though it continued to be published annually. Information about Ordnance Survey publications is contained in the monthly *Publication Report* in which new and revised editions of maps at all scales are listed. The Ordnance Survey also issues *Map News and Review* about eight times per year to provide information on future publications for trade and educational customers.

The number of periodicals and serial publications which are devoted entirely to cartography, or may contain articles or reviews on cartographical material, is now enormous. Selected geographical and cartographical serial publications that contain lists and/or reviews of current maps and atlases are listed in the *Bulletin* of the Special Libraries Association, Geography and Map Division, No. 102 (December 1975). In an excellent review of periodical and serial publications K. A. Salichtchev of Moscow State University (1979) quotes the annual bibliography of cartographic literature *Bibliographia Cartographica* (volume 3, 1976) which listed no fewer than 188 such publications. Of these only twenty were strictly cartographic journals; another twenty were cartographic–geodetic and fifty-eight were geographic. Salichtchev, however, points out that even a total of 188 is an underestimation, noting that the journal *Kartographische Nachrichten* in a survey of periodicals for 1976 includes a number not cited in *Bibliographia Cartographica*. The most recent issues of this journal include citations from nearly 300 periodicals but the proportions given in Salichtchev's review remain fairly similar. However, with more than 2,000 citations, *Bibliographia Cartographica* remains the most comprehensive bibliography of cartography [336].

Several international organizations issue periodicals concerned with maps and mapmaking, the two outstanding examples being *World Cartography* [401], published since 1951 by the Cartographic Office of the Department of Social Affairs of the United Nations, and the *International Yearbook of Cartography* [389], issued since 1961 under the auspices of the International Cartographic Association. *World Cartography* was conceived as an annual publication but in that respect has not quite lived up to expectations, for several years have gone by without its appearance. Nevertheless, it remains an important reference source for the mapping of the world from official data and also for the exchange of knowledge on particular problems of mapmaking, some of which may be especially significant for developing nations. The *International Yearbook* also stands for the interchange of cartographical achievements and knowledge, containing as it does articles on every aspect of topographic and thematic mapmaking, including some interesting light on the history of major commercial firms and organizations.

As far as national publications are concerned the United Kingdom has been well to the fore. The premier publication, *The Cartographic Journal* [381], is the organ of the British Cartographic Society and has been published twice-yearly since the formation of the society in 1964. The diversity of membership of the Society—professional cartographers with both official and commercial bodies, free-lance workers, academics, cartographers in higher education, map curators and so on—is reflected in the range of topics covered though it is noticeable that a large number of the papers have emanated from academic sources. The history of cartography is well represented, as is the current state of cartographic production and technology. Reviews, news items, lists of new maps and publications are also featured. Less ambitiously produced, but with a content that is greatly to the credit of a small society, the Society of University Cartographers, is the *SUC Bulletin* [377], published twice-yearly at slightly irregular intervals since 1964. This journal contains scholarly articles on every aspect of cartography but is especially notable for the size and quality of its review section, which covers equipment as well as literature. The other major British journals concerned with mapping are *The Photogrammetric Record* [433] published twice yearly by The Photogrammetry Society, and the *Survey Review* [399], a quarterly publication issued by the Commonwealth Association of

Surveying and Land Economy and largely devoted to technical articles on photogrammetry and surveying.

*The Geographical Journal*, the quarterly journal of the Royal Geographical Society, occasionally features articles concerned with the history of cartography and with current developments, while the *Geographical Magazine*, with a very wide readership, occasionally reviews books on cartography and has sometimes devoted whole issues of the magazine to cartography, examples being those of April 1960, 'Atlases and mapmaking' or October 1969, 'Cartography for the 1970s'.

Several cartographic journals are published in Europe, the oldest-established being *Kartographische Nachrichten* [390], published bimonthly by the Cartographic Society of the Federal Republic of Germany. This journal appeared first in 1951 but since 1976 has become international as the journal also of the Cartographic Society of Switzerland and the Cartographic Commission of the Austrian Geographic Society. All aspects of cartography are covered but it is specially important for its coverage of the present state of mapmaking in the three countries.

The Dutch contribution to cartographic development, both historically and in modern times, is out of all proportion to the small size of the country. The Dutch Cartographic Society published a quarterly journal, the *Kartografisch Tijdschrift* [391], which includes articles and reviews in Dutch on the history of cartography in the Netherlands, modern technology and methodology.

The Italian Cartographic Society (Associazione Italiana di Cartografia) has, since 1964, published its *Bolletino della Associazione Italiana di Cartografia* [373] which, in addition to the usual papers on the history of cartography and modern mapmaking, once a year includes a list of the publications of official and private cartographers. Abstracts in English are provided.

The Spanish *Boletín de Información* [Information Bulletin] [372] of the Servicio Geográfico del Ejército (Geographical Department of the Army) is published quarterly in Madrid. Its emphasis is on topographic mapping of Spain but modern technology elsewhere is also covered. An earlier journal, *Boletín de Cartografía* [Bulletin of Cartography], founded in 1961, was short-lived and failed to remain economically viable. The French *Bulletin du Comité Français de Cartographie* [376] has maintained an irregular schedule since 1958 and covers a wide range of topics. The occasional issues devoted to a single theme are particularly interesting, as for example the *Explanatory Dictionary of French terminology in cartography* (No. 64, 1970). *Mitteilungen der Österreichischen Geographischen Gesellschaft* [Reports of the Austrian Geographic Society] has contained articles on the history of cartography and current mapping since 1949 and was joined in 1976 by the already mentioned *Kartographische Nachrichten* as the official journal of the Cartographic Commission of the Austrian Geographic Society.

The Swedish quarterly, *Globen* [387], may be regarded as the pioneer of all cartographic journals as it has appeared since 1922 though its function is a little different from that of the publications of most cartographic societies. To some extent it may be regarded as an advertising medium for it is published by Esselte and is largely concerned with the topographic maps of Sweden, and particularly with Esselte publications.

Of the European socialist countries Hungary has been foremost in dispensing cartographic information, both by international conferences and by publishing of such periodicals as *Cartactual* [378]. The latter, with six issues per annum, is a unique publication of immense value to all whose concern is with the preparation of maps for it provides up-to-date information in map form on boundary changes, communication networks, the establishment of power-stations, changes in place names and so on. Since 1971 it has been supplemented by a further publication, *Cartinform* [379], a serialized bibliography of current cartographic literature, maps and atlases. Poland has two journals featuring cartography: *Geodezja i Kartografia* [Geodesy and Cartography] [385] has been published quarterly by the Geodetic Committee and the Polish Academy of Sciences and is biased towards geodesy; *Polski Przeglad Kartograficzny* [Polish Cartographical Review] [396] is also a quarterly and has been published since 1939 by the Polish Geographical Society's Cartographic Commission and the State Map Publishing Establishment. This is a high-quality journal with scholarly contributions from Polish cartographers, as well as reviews and sections dealing with current Polish maps, atlases and literature. *Geodeticky a Kartograficky Obzor* [Geodetic and Cartographic Review] [383] is the monthly journal of the Czech and Slovak Directorate of Geodesy and Cartography and has been in existence since 1955.

The main cartographic journal in the U.S.S.R. is *Geodeziia i Kartografiia* [Geodesy and Cartography] [384], a monthly journal which is the organ of the Central Directorate of Geodesy and Cartography of the Council of Ministers of the U.S.S.R. Salichtchev tells us that it was established under the name *Geodezist* [Geodesist] in 1925 and was the first of the world's carto-geodetic journals. The emphasis is strongly on geodesy with only about two articles per issue on purely cartographic matters. Specialized themes, however, are occasionally dealt with in the appropriate professional journals, for example geological mapping in *Sovetskaia Geologiia* [Soviet Geology] and soil mapping in *Pochvovedenie* [Soil Science]. A publication of major interest is *Geobotanicheskoe Kartografirovanie*, which is issued by the V. L. Komarov Botanical Institute and is a unique example of a specialized journal entirely devoted to one specific mapmaking theme. The Book Bureau publishes *Kartograficheskaia Letopis* [Cartographic Yearbook], an annual bibliography of Soviet maps.

The Canadian journal now published under the title *Cartographica* [380] has had an interesting history and is unique in some respects. It came into existence in 1965 as a twice-yearly publication entitled *The Cartographer*, emanating from York University, Toronto, under the editorship of Dr. B. V. Gutsell, who has done so much for Canadian cartography. In 1968 the name was changed to *The Canadian Cartographer* and more recently to the present title, *Cartographica*. The standard of production has increased over the years and is now of the highest quality. The content consists of a good blend of papers devoted both to historical themes and modern techniques. The amount of coverage given to Russian cartography, particularly the history of Russian mapmaking, is unequalled in any journal published in the West. Indeed Salichtchev states that in this respect it can compete with any Soviet journal. A unique feature is the number of issues devoted to a single theme; a series of such volumes appeared under the general title of Cartographica before the journal itself adopted the title. Such themes have included 'Map design and the map user'; 'Eskimo maps from the Canadian eastern Arctic'; 'Cartographic generalization'; 'The seven aspects of a general map projection'; 'Computer cartography in Sweden'; and 'The nature of cartographic communication'.

In the United States of America the American Congress on Surveying and Mapping has issued a carto-geodetic journal, *Surveying and Mapping* [400], since 1941. The content reflects the interests of the membership of the Congress: land and cadastral surveying; civil engineering; geodetic survey; cartography. The cartographic content is generally related to the production of topographic maps. In 1974 a second journal, *The American Cartographer* [370], appeared under the auspices of the Congress of Surveying and Mapping, and under the distinguished editorship of Professor Arthur Robinson, professor of cartography at the University of Wisconsin, the outstanding establishment for cartographic studies in the United States. *The American Cartographer* is a twice-yearly journal on similar lines to its fellows published elsewhere and is particularly good for its coverage of the theoretical side of cartography, computer-aided mapping and educational matters.

The year 1970 saw the publication of the first issue of *Revista Brasileira de Cartografia* [The Brazilian Journal of Cartography] [397], a journal which, after irregular appearances during its first three years, has now settled down to two issues per annum. *Revista Cartografica* [the Cartographic Journal] [398] has been published by the Military Geographic Institute in Argentina since 1951.

*Map* [392] is the quarterly organ of the Association of Japanese Cartographers and has been published since 1963 with Japanese text and abstracts in English of the main contributions. Once again this journal fulfils a useful purpose both internally and externally as a reference source of information about Japanese cartographic practice and production, while a newcomer to the Asian scene is *Indian Cartographer*, journal of the Indian National Cartographic Association.

*Cartography* [382] is the biannual publication of the Australian Institute of Cartographers and has been issued from Canberra since 1954 with a bias towards modern production and technology as well as geodesy.

The *South African Journal of Photogrammetry, Remote Sensing and Cartography* (formerly *The South African Journal of Photogrammetry*) [434] is published by the South African Society for Photogrammetry, Remote Sensing and Cartography with financial support from the Department of National Education. Its articles and reviews are considerably biased towards photogrammetry and remote sensing but they also have a reasonable amount of cartographic coverage.

Many other serials provide information about cartography in the form of reviews and lists of new publications. One of the most useful was *New Geographical Literature and Maps* [394], a biannual publication of the Royal Geographical Society in London. Its main purpose was to list new material that had been added to the Society's collections, and a section headed 'New atlases and maps' identified all the current accessions to the Society's Map Room in Kensington Gore. Regrettably, the Society has been unable to sustain the publication.

*Current Geographical Publications* is a similar list and is published ten times per year by the American Geographical Society. It covers accessions concerned with geography, not merely maps, and is divided into three sections: General; Regional; and Maps. The Bodleian Library in Oxford offers a monthly listing of *Selected Map and Book Accessions* and the British Library updates its important *Catalogue of printed maps* with listings in *Accessions*.

Apart from major national collections, several university map libraries prepare lists showing acquisitions. Lists of such libraries and other institutions that prepare acquisition lists are occasionally given in the *Bulletin* [375] of the Special Libraries Association, Geography and Map Division. Donald A. Wise (1978), in one of a series of appendices under the general title 'Cartographic sources and procurement problems', included lists of selected serials containing lists and/or reviews of current maps and atlases and also the titles of selected map and atlas accessions lists published by organizations throughout the world.

Carto-bibliography and the history of cartography have their own specialized journals as well as a vast array of literature in book form, selected examples of which were noted in Chapter 2. The major journal is *Imago Mundi* [34], the organ of The International Society for the History of Cartography. Originally published in the Netherlands, *Imago Mundi* is now edited by Professor Eila Campbell and contains important contributions to the study of early cartography. A second British journal concerned with the history of cartography, *The Map Collector* [327], is primarily concerned with the needs of map collectors. Published by Map Collector Publications of Tring, Hertfordshire, *The Map Collector* is a quarterly journal, lavishly produced and with an impressive range of contributors. The first issue appeared in 1977 and among its many features is a continuing list of dealers in antiquarian maps, probably the most useful and comprehensive to be found in any source.

## Research into Cartography and Cartographic Development: Dissertations

Dissertations can be an important information source and there are occasions when it is necessary to be aware of the state of contemporary research into a particular topic—to check whether one's own research might be duplicating another's work, for example. The *International directory of current research in the history of cartography and in carto-bibliography* [32] appeared for the fourth time on the occasion of the 10th Conference on the History of Cartography held in Dublin in

1983. This is the major reference source into current work in the fields of carto-bibliography and the history of cartography.

Sources of information on dissertations are listed by Chauncy D. Harris (1976) as part of a bibliography of geography. Dissertation lists are generally devoted to those of a single country and cartography is likely to be found among the geography listings. In Britain the Royal Geographical Society's publication *New Geographical Literature and Maps* [394] provided annual lists of completed theses, and theses (including those in preparation) are also listed in volumes issued annually by the Conference of Heads of Geography, United Kingdom and Ireland. In the United States the journal *Professional Geographer*, published in Washington, D.C. by the Association of American Geographers, features lists of recent geography theses including those in preparation. For Canada, Fraser and Hynes (1966) listed dissertations concerned with Canadian geography and in 1972 Fraser compiled a supplement to this listing. Marsden and Tugby (1971) compiled a preliminary edition of Australian geography theses published between 1933 and 1971.

One of the main advantages of dissertations as a source of cartographic information is that the topics covered are generally original, the author having been likely to ensure that his research has not already been covered by some earlier researcher.

## Aerial Photographs

Aerial photographs are of considerable potential interest to the cartographer, as indeed is ground photography. Such photographs are widely used in conjunction with maps in the study of land use, soils, technics and so on. One of the most useful sources of information about photographic material is the *High School Geography Project. Sources of information and materials; maps and aerial photographs* (1970). This was an ambitious project covering books, papers and other sources of information (mainly American) on ground, aerial and space photographs; maps and atlases. The chapters include 'Annotated bibliography on maps and mapping, aerial photography and remote sensing'; 'Filmstrips, slides and films'; 'A selected list of statistical data sources', 'Matching topographic maps and aerial photographs'; 'Sheet maps and other cartographic materials'; 'Maps, globes, relief models, transparencies of maps from commercial sources; list of map agents by states'; 'Aerial and space photography with sources for obtaining aerial photographs of the United States'; 'Equipment for use in map and aerial photographic interpretation; materials and equipment used in map preparation'; and 'List of addresses of commercial firms'. A project of this kind is clearly a valuable source of information not only on sources of aerial photographs but on a variety of other cartographic topics also.

Other sources of information are the catalogues of commercial firms, whose advertisements are sometimes to be found in the various cartographic journals and the national survey organizations.

## Bibliographies of Geography

Cartography is one of the subjects included in several bibliographies of geography and general reference works, some of which were described and listed by Chauncy D. Harris (1970). The best of these general bibliographical works is that by Wright and Platt (1971), a very comprehensive work that lists 1,174 items and includes sections on bibliographical aids to maps and cartography; gazetteers and world atlases. C. B. Muriel Lock's *Geography: a reference handbook* includes among its 1,283 entries items on the history of cartography, atlases and maps, together with biographical notes on individual cartographers, while Stephen Goddards' *A guide to information sources in the geographical sciences* includes a chapter on 'Maps, atlases and gazetteers' by G. R. P. Lawrence, and one on 'Aerial photographs and satellite information' by J. A. Allan.

Information on Japanese cartography was provided by Hall and Noh (1970), whose work includes sections on yearbooks; statistics and censuses; atlases; maps; aerial photography. In all Chauncy Harris listed twenty-two geographical bibliographies from Great Britain, the United States, Canada, Germany, the U.S.S.R., Czechoslovakia, Italy and Japan. He also listed and annotated important serialized bibliographies of geographical material which normally include information on maps and cartography. For example, *Bibliographie Géographique Internationale* has been published in France since 1891 and, as a result of widespread international co-operation from agencies as well as individuals, offers the most comprehensive coverage of the geographical field. *Current Geographical Publications* commenced publication in the United States in 1938 with ten issues per annum. Part 1, Section 39, is concerned with cartography and Part 3 is given over to maps. *Referativnyi Zhurnal: Geografiia* [Reference Journal: Geography] has been published twelve times per year in Moscow since 1954, its series M being devoted to cartography. The English series, *GEO Abstracts*, is published from the University of East Anglia at Norwich and features seven separate series, each with six issues per year. Of these, Series G is concerned with remote sensing and cartography. *Documentatio Geographica* is published by Herausgegeben von Bundesanstalt für Landeskunde und Raumordnung, Bonn–Bad Godesberg and includes a small number of entries on cartographical matters.

Retrospective bibliographical material in geography is found in three major sources: the *Bibliographie Géographique Internationale*, which has already been mentioned; the *Geographisches Jahrbuch*, published by the famous firm of VEB Hermann Haack Geographisch-Kartographische Anstalt, Gotha/Leipzig; and the *Research Catalogue of the American Geographical Society*, published by G. K. Hall & Co. of Boston, Massachusetts. All three publications contain some citations relating to cartography.

## Record Societies

The transactions of county and local record societies often include material of cartographic interest as well as occasionally devoting whole issues to a particularly significant local map. An outstanding British example is J. B. Harley's work on Christopher Greenwood's map of Worcester, which was published by the Worcestershire Historical Society in 1962 and which includes much valuable information about Greenwood's mapmaking business, the way in which his maps were made, and the relationship between the map of Worcestershire and the landscape of the county. The Essex Record Office in Chelmsford has been to the fore in publishing books on the early mapping of the county; F. G. Emmison edited the *Catalogue of maps in the Essex Record Office, 1566–1860* and the introduction to this catalogue formed the text of a well-illustrated publication, *The art of the map-maker in Essex* (1947). This is a particularly significant work in that the Essex Record Office has a fine collection of old manuscript maps of estates and farms. Formerly dispersed in country houses, estate offices and with solicitors, these maps are now brought together to provide an unusually rich field for research. The Lancashire and Cheshire Historic Society published a monograph on *William Yates's Map of Lancashire 1786* (1968) which included an important introduction by J. B. Harley, and in volume 1 of an Occasional Series the same society issued *A survey of the County Palatine of Chester. P. P. Burdett. 1777* (1974) with a full-scale facsimile of Burdett's map and an introduction by J. B. Harley and P. Laxton that gave valuable insights into the theory and practice of eighteenth-century county surveying.

Similar works have appeared elsewhere. For example, Ernest J. Burrus wrote *Kino and the cartography of northwestern New Spain* (1965), a work that described the life, work and principal cartographic contributions of Eusebio Francisco Kino. Edgar Crosby Smith edited *Moses Greenleaf: Maine's first map-maker. A biography: with letters, unpublished manuscripts and a reprint of Mr. Greenleaf's rare paper on Indian place-names: also a bibliography of the maps of Maine* (1902), a work which, like so many of its kind, furnishes the researcher with information about little-known figures in cartographic development that is not readily accessible elsewhere.

## Topography

The publications of the London Topographical Society are of great interest to the cartographic historian and to the urban historian. Many of this Society's publications are of maps, plans and views that illustrate the history and topography of London, and the annual volumes of the Society (which have a general title, the *London Topographical Record*) contain articles relating to London maps and plans. Examples of plans issued and described by the Society include Horwood's famous plan of 1792–99, Thomas Milne's land-use map of London and environs in 1800, and detailed plans showing the seventeenth-century properties of St. Bartholomew's Hospital.

## Exploration and Discovery

While exploration is not strictly speaking part of cartography, it is important in the data acquisition process; indeed, for centuries it was the main source of data for mapmaking. A good example of its importance is seen in the Ptolemy-derived maps of the Renaissance, which were centuries out of date and had to be revised to accommodate new material brought in by the great explorers of the day. R. A. Skelton provided the link between exploration and mapmaking in his *Explorers' maps: chapters in the cartographic record of geographical discovery* [68], a work that not only relates the exploits of famous explorers like Columbus, Balboa (the first European to set eyes on the Pacific Ocean), Frobisher, Livingstone—who, while critically ill in Central Africa, continued to record the astronomical observations and calculations by which he determined his latitude and longitude—but also shows how maps and charts developed from data provided by medieval travellers and how explorer/surveyors such as Captain Cook were able to prepare scientifically constructed charts. Skelton has provided historians with a considerable service by reproducing over two hundred important maps and engravings dating back in some cases to 1375. Skelton also contributed to *The discovery of North America* (1971), a work which, with its companion volume *The exploration of North America 1630–1776* (1974), provides a magnificently illustrated account of the early explorations and of the multi-national expansion of colonial North America. Each volume is a splendid source of reproductions of early maps, world as well as regional, many of which are superbly reproduced in colour. The same can often be said of works devoted to individual explorers and aimed at a general readership. One example is *The voyages of Columbus* (1970) by Rex and Thea Reinits, a volume that has a plethora of illustrations of early maps, engravings, artefacts, botanical and biological drawings.

The Hakluyt Society [1397], established in 1846, has published nearly three hundred volumes of original narratives of voyages, travels, expeditions and other geographical records. The volumes are issued free to members of the Society or may be purchased by non-members through a bookseller or the Society's distribution agent.

## References

Burrus, Ernest J. *Kino and the cartography of northwestern New Spain.* Tucson: Arizona Pioneers' Historical Society, 1965. 104pp.

Cumming, W. P., S. Hillier, D. B. Quinn and G. Williams. *The exploration of North America 1630–1776.* London: Paul Elek, 1974. 272pp.

Emmison, F. G. *Catalogue of maps in the Essex Record Office 1566–1860.* Chelmsford: Essex County Council, 1947. 126 pp.

Emmison, F. G. *The art of the map-maker in Essex 1566–1860.* Chelmsford: Essex County Council, 1947. xvpp.

Fraser, Keith J. and Mary C. Hines. *List of theses and dissertations on Canadian geography.* Department of Mines and Technological Surveys, Geographical Branch, 1966. (Bibliographical Series 31.)

Goddard, S. (ed.) *A guide to information sources in the geographical sciences.* Beckenham: Croom Helm, 1984. 272pp.

Hall, Robert B. and Toshio Noh. *Japanese geography: a guide to Japanese reference and research materials*. Revised edn. Ann Arbor, Michigan: University of Michigan Press, for the Center for Japanese Studies, 1970.

Harley, J. B. *Christopher Greenwood, county map-maker and his Worcestershire map of 1822*. Worcester: Worcestershire Historical Society, 1962. 72pp.

Harley, J. B. *William Yates's map of Lancashire 1786*. The Historic Society of Lancashire and Cheshire, 1968. 48pp.

Harley, J. B. and P. Laxton. *A survey of the County Palatine of Chester. P. P. Burdett. 1777*. The Historic Society of Lancashire and Cheshire Occasional Series, vol. 1 (1974). 37pp.

Harris, Chauncey D. *Bibliography of geography*. Part 1, *Introduction to general aids*. Chicago: University of Chicago, 1976. (Department of Geography Research Paper 179.) 276pp.

High School Geography Project. *High School Geography Project. Sources of information and materials; maps and aerial photographs*. Association of American Geographers, Committee on Maps and Aerial Photographs, 1970.

Lock, C. B. Muriel. *Geography: a reference handbook*. 3rd edn. London: Clive Bingley; Hamden, Connecticut: Linnet Books, 1972. 762pp.

Marsden, B. S. and E. E. Tugby. *Bibliography of Australian geography theses*. University of Queensland, Department of Geography, 1971. 77 pp.

Reinits, Rex and Thea Reinits. *The voyages of Columbus*. London, New York, Sydney, Toronto: Hamlyn, 1970. 152pp.

Salichtchev, K. A. 'Periodical and serial publications on cartography'. *The Canadian Cartographer*, **16**(2): 109–132 (1979).

Skelton, R. A., W. P. Cumming and D. B. Quinn. *The discovery of North America*. London: Paul Elek, 1971. 304pp.

Smith, Edgar Crosby (ed.) *Moses Greenleaf: Maine's first map-maker. A biography: with letters, unpublished manuscripts and a reprint of Mr. Greenleaf's rare paper on Indian place-names: also a bibliography of the maps of Maine*. Bangor, Maine: C. H. Glass & Co., 1902. 192 pp.

Wise, Donald A. 'Cartographic sources and procurement problems. Appendix A. Selected serials containing lists and/or reviews of current maps and atlases'. Special Libraries Association, Geography and Map Division, *Bulletin*, **112**: 19–22 (June 1978). Appendix B. 'Selected map and atlas accession lists'. *Bulletin*, **112**: 23–26 (June 1978).

Wright, John Kirtland and Elizabeth T. Platt. *Aids to geographical research: bibliographies, periodicals, atlases, gazetteers and other reference books*. Westport, Connecticut: Greenwood Press, 1971. 331pp.

# 5 Keeping Up to date with Current and Forthcoming Events, Publications and Developments

In the post-war years enormous advances have been made in the technology of cartography and these have been paralleled by an impressive increase in the range of events that regularly take place in the mapmaking world. The diversity of current events and happenings includes technical workshops, symposia, summer schools, one-day meetings, educational programmes, visits to venues of cartographical interest and so on. Such events are normally limited to members of the organizing society but there are occasions when circumstances allow the participation of non-members who have a serious interest in the topics being covered. How does one find out about future programmes, where the venues are to be, and how does one apply to attend? Fortunately there are several sources to which would-be participants can turn for news and also for detailed reports about similar events that have already taken place. The most important and most up to date of such sources are the newsletters published by the various cartographic societies and associations but usually distributed to members only. The next best source is the news columns of specialized cartographic journals such as *The Cartographic Journal* [381], the main drawback being that such journals are often published only twice or four times per year and consequently news items can occasionally appear a little late.

## *Newsletters*

Undoubtedly the most up-to-date and useful guides to future events are the newsletters issued by the various professional societies, libraries, official cartographic agencies and commercial firms. They are less lavishly produced than the major journals, and the simpler processes used to print the newsletters mean that readers can be informed in good time about future items and happenings of cartographic interest.

Most cartographic societies issue newsletters to their members, normally twice per year. The *Newsletter* [406] put out by the British Cartographic Society is a typical example of the genre. Currently edited by David Fairbairn of the Department of Surveying, University of Newcastle upon Tyne, it consists of twelve A5-size pages in which are featured editorial columns; reports of meetings and lectures; a series of company profiles, which are of considerable interest (the *Newsletter* No. 2 for 1984 looks at the Ordnance Survey, George Philip & Son Ltd., The Automobile Association and Agfa-Gevaert); and a very useful innovation which provides information not readily available elsewhere: news of forthcoming workshops, symposia, lectures and seminars.

One of the most recently established newsletters is that of the International Cartographic Association of which the first issue was dated June 1983. The ICA's *Newsletter* [408] is to be published twice per year and is edited jointly by Roger Anson in the United Kingdom and Bernard Gutsell (well known for his work in producing *Cartographica*) in Canada. The first number began with a report on the 11th International Cartographic Conference held in Warsaw in 1982 and this was followed by reports from some of the ICA special commissions which were set up in Tokyo for the period 1980–84. Some of these commissions have already published material on their findings: the Commission on Cartographic Technology, for example, was responsible for the publication of *Colour proofing systems in cartography* (1980), and future publications are to include *A basic manual of cartography* and a new edition of the *Multilingual dictionary of technical terms in cartography* [367]. Reports also appear in this newsletter on matters for which the ICA is not responsible; the first issue, for instance, reminds us that the Nederlands Kartografische Vereeniging [Netherlands Cartographic Society] celebrated its twenty-fifth anniversary in 1983.

Newsletters issued by the numerous professional societies and associations generally follow the pattern of the British Cartographic Society. The Society of University Cartographers, another British organization but one with worldwide membership, was one of the very first bodies to issue a newsletter. Presently edited by Steve Chilton of the School of Geography and Planning, Middlesex Polytechnic, the *SUC Newsletter* [405] is a twice-yearly publication that is designed primarily for cartographers who are employed in establishments of higher education. It contains news of forthcoming summer schools held by the Society, details of conferences promoted by other organizations, news of exhibitions, minutes of the Society's annual general meetings and a particularly useful item: job vacancies.

Specialized societies also issue their own newsletters. *Sheetlines* [409], the information sheet of the Charles Close Society, is a case in point, being largely concerned with the history of the Ordnance Survey. The Remote Sensing Society has as one of its stated aims and services the provision of newsletters and remote sensing letters and also provides an important information service in that it encourages those with scientific or commercial enquiries on remote sensing topics to get in touch with the Society's secretary, who will then put them in touch with a competent specialist, a relevant institution or the appropriate trade association. This seems a remarkably enlightened policy and one that other cartographic societies might follow.

Several bodies in the United States of America issue useful information sheets and newsletters. The official mapmaking agency, the U.S. Geological Survey, issues a quarterly *National Cartographic Information Center Newsletter* [407], which is available from NCIC, USGS, Mailstop 507, National Center, Reston, Virginia 22022. The primary objective of this newssheet is the provision of information about the Center, where a database is being developed which will store information on what cartographic data are held where and by whom. The Center also aims to set up systems by which such information can be distributed as rapidly as possible whenever it is required. The American Library Association's Map and Geography Round Table (MAGERT) also issues a quarterly newsletter, *base line* [402], which provides news of recent publications in any format relating to cartography, maps and geography, as well as details of conferences and meetings. *base line* is also concerned with more specialized topics such as map cataloguing. *Mapline* [403] is the quarterly newsletter of the Hermon Dunlap Smith Center for the History of Cartography at the Newberry Library in Chicago. In addition to news items from map societies and major map collections, news of exhibitions, details of recent publications and a calendar of future events, *Mapline* generally features a major article on a topic relating to the history of cartography.

## *Professional Journals*

The journals issued by cartographic societies and others are also a prime source of information on forthcoming events, new equipment, new techniques and new publications as well as containing reports on those events that have already taken place at the time of publication. *The Cartographic Journal* [381], for example, annually contains a Society Record in which the president of the British Cartographic Society reviews the events of the previous year. It also includes an international diary of events that are to take place throughout the world together with details of where to write to obtain information about each event. The section headed 'Recent maps and atlases' is compiled by Betty D. Fathers from material received in the Map Section of the Bodleian Library, and is a good pointer to the most important recently published maps and atlases. In the same journal Michael Wood provides a list of recent literature. The June 1984 issue of *The Cartographic Journal* (volume 21, No. 1) contained a particularly important review of the cartographic activities in the United Kingdom from 1980 to 1983. This report had been prepared on behalf of the Cartography Subcommittee of the Royal Society's British National Committee for Geography and submitted to the 7th General Assembly of the International Cartographic Association.

Readers in the United States have the benefit of that prime information source, the *Bulletin* [375] of the Special Libraries Association, Geography and Map Division. While this journal is designed as a medium of exchange of information and research in the field of geographic and cartographic bibliography, it also features several sections that enable the map curator, the librarian and the general map user to keep in touch with current events, new literature and new cartographic products. From time to time lists are provided of publishers of a variety of material. For example, in *Bulletin* No. 128, June 1982, Donald A. Wise compiled 'A selective list of international publishers and distributors of globes and three-dimensional plastic relief models (maps)' (pp. 36–38) and in rather different vein 'A list of national bibliographies and references containing citations on atlases and maps' (pp. 39–41). In the same issue Charles A. Seavey contributed a feature entitled 'New government publications of interest'. Another useful feature that appears frequently is a list of technical reports in geography and mapping, and the *Bulletin* is also one of the very few to provide a list of members of the Association.

*IMCØS Journal* [328], the quarterly publication of the International Map Collectors' Society, is a comparative newcomer to the scene and, although fairly modestly produced, it is a useful guide to events in the map collecting field. It contains editorial news and views, information about future meetings of IMCØS, reports on IMCØS meetings and conferences, articles on the history of cartography and a 'library report' which gives details of new acquisitions by the Society's library.

## Reviewing Media

The professional journals are naturally the best source for reviews of new literature on cartography, new maps and atlases, and in some cases of new equipment. The major drawback as far as keeping up to date is concerned is that a considerable span of time may elapse between the publication of a book or atlas and the appearance of its review. Delays of this kind are a major headache for review editors and the problem seems an insoluble one given that most journals appear only on a twice-yearly basis. Nevertheless the scholarly reviews that appear in almost every cartographic journal are an important guide to the suitability and quality of new literature and products.

The most extensive of all the review sections is that in the Society of University Cartographers *SUC Bulletin* [377]. This journal carries an average of fifteen pages of book reviews and map and atlas reviews in a section edited by Terry Garfield of the University of Leicester. From time to time the *Bulletin* also includes news and reviews of equipment, particularly the type of equipment used in small cartographic units. Specialized journals such as the *South African Journal of Photogrammetry, Remote Sensing and Cartography* [434] often carry reviews of general cartographic literature as well as more specific remote sensing material.

Cartographic journals are by no means the only media providing reviews of maps, atlases, cartographical products generally and literature concerned with cartography. Many of the geographical serial publications occasionally feature reviews and news items of cartographical interest. Among them, the American *Geographical Review* and *Professional Geographer*, the British *Geographical Journal*, *Geographical Magazine*, *Geography* and the *Transactions* of the Institute of British Geographers are particularly useful sources. As already mentioned in the previous chapter, Donald A. Wise has listed serials containing lists and/or reviews of current maps and atlases, etc. in the *Bulletin* of the Special Libraries Association, Geography and Map Division.

Most journals publish annual indexes in which information about reviews on a particular topic can be sought.

## Symposia

Several contemporary cartographic societies and organizations hold regular conferences, symposia, summer schools, technical workshops, one-day courses and lecture meetings. While these are often open only to members of the specific society, interested persons and serious students are occasionally accommodated. Most of these events, except for the technical workshops where specific practical projects are dealt with, follow a similar pattern, with a programme consisting of a series of technical lectures and seminars; exhibitions of cartographic materials and equipment; exhibitions of new books, maps and atlases; exhibitions of members' work; visits to important map collections or mapmaking establishments. Some associations provide a useful service for those unable to attend their meetings by publishing the manuscripts of the lectures given at the annual symposium. The *Publications* of the Remote Sensing Society are probably the best example of this type of publication. Other societies provide a summary of previous lectures in their journal.

## Accession or Acquisition Lists

The librarian or map curator who wishes to keep abreast of the tide of current cartographical material and literature can be kept up to date to some extent by having himself placed on the mailing lists for distribution of the periodic accession or acquisition lists that are compiled and issued by many map collections and libraries.

Typically such lists are an itemization of materials added to a collection over some given period and, while they are often intended primarily to inform internal staff and local users of new acquisitions, they also act as a useful communications medium and as an acquisition tool for curators and librarians elsewhere. The *Bulletin* [375] of the Special Libraries Association, Geography and Map Division is once again extremely helpful in this field. *Bulletin* No. 112 for June 1978 carries on pp. 23–26 a very helpful list of selected map and atlas acquisition lists, and another useful paper, 'Preparing acquisition lists', written by Lewis A. Armstrong and Margaret T. Pearce, appeared in issue No. 104 of the same journal, dated June 1976, pp. 34–37. The substance of this latter article was derived from answers to a questionnaire sent out to the keepers of all major map collections in the United States during August 1974. The questionnaire requested information on purpose, content, preparation and use of acquisition lists. The replies received suggested that information would normally be given according to the following priorities: topical subject maps; general maps; topographic maps; and oil company road maps. For each item the information provided would include title, publisher with address, the map's scale and its date of publication. Further useful information that might be given could indicate whether the map was printed in colour or monochrome; whether it was a single sheet or formed part of a series; the dimensions of the map; its library reference number; whether it was a flat or folded sheet. Curators of map collections may find it of interest to read that the average total time allocated to publishing an acquisition list was only six hours. This does not seem a great deal of time when such a useful service can be initiated, for the generally wide distribution of acquisition lists emphasizes that they do form an important communication link between curators of map collections and also serve as a helpful tool in assisting curators in the selection of significant material with which to bolster their own collections.

In view of the immense size of its map collection it is perhaps hardly surprising that the Library of Congress, Geography and Map Division issues one of the largest and most ambitious of acquisition lists. This forty-eight-page publication covers cartographic items received by the Library and it is interesting to note that during 1981 alone some ninety-four thousand items were added to the permanent collections of over three and a half million maps. The list is arranged in the following categories: Americana; foreign cartography; cartographic miscellany; Library of Congress transfers; current printed maps and atlases; and selected acquisitions. This important reference source can be obtained free of charge from the Library of Congress, Geography and Map Division, Washington, D.C. 20540.

## Services

The major map collections, some of which have holdings running into many thousands, indeed in some cases millions, of sheet maps, atlases, globes and models, house such a vast amount and range of material that the potential user may need some guidance to help him familiarize himself with the content and arrangement of the collections and the services that may be offered. In the United States the collections of the Library of Congress Geography and Map Division have increased from a remarkably modest three maps and four atlases at the time of inception in 1800 to the present three and a half million sheet maps and thirty-eight thousand atlases. The Library issues a guide that introduces the reader to its special collections, its materials from different historical periods, includes sections on different types of map and chart and concludes with information about the services and organization of the Geography and Map Division. Similar assistance is provided by many other keepers of map collections.

A service of a rather different kind is provided by the Map Library of the University of Illinois, which makes available to all map users a sizeable worldwide listing of map sources. Stored on some ten thousand cards, the information is broadly classified into three categories: United States sources; colleges and universities; and other countries. Within each section the information is broken down into a great number of different types of organization having holdings of maps and cartographic material. A similar listing created by the York University Library (Ontario) is published as the *Map sources directory* [631].

## Exhibition Catalogues

Many of the major institutions housing map collections mount occasional exhibitions of cartographic material, which perform an important educational function, not only in allowing serious students of cartography to have the opportunity to study a wide range of products, including rare and historically significant maps and atlases, but also in bringing the fascination of maps and mapmaking to the notice of members of the general public, whose previous awareness of maps may, in many cases, extend no further than a 1:50,000 Ordnance Survey sheet or a motoring map.

The staff of the Map Room of the British Library in London have been to the fore in arranging superb exhibitions drawn from the vast range of items under their care. Topics covered have included 'Christopher Saxton and Tudor Map-Making'; 'Willian Roy, Pioneer of the Ordnance Survey'; a fascinating exhibition of cartographical curiosities; and Chinese and Japanese maps, a field from which examples are not very often seen in the Western world. One of the most useful aspects of these exhibitions is the publication of scholarly descriptive catalogues, usually well illustrated and in some cases the only publication to deal with a particular topic. Undoubtedly the most lavish and comprehensive of all map exhibition catalogues was *Cartes et figures de la Terre* [349], a superbly produced 480-page volume, illustrated in colour and monochrome, which was published to accord with a wide-ranging exhibition of maps and other cartographical products at the Centre Georges Pompidou in Paris in 1980. Even though the text is in French only, *Cartes et figures de la Terre* is a mine of information on topics as widely disparate as the Madaba Mosaic and modern maps in advertising. A useful glossary of technical terms is included.

Highly significant exhibitions have been organized by the Amsterdam Historical Museum. Cartographically speaking, the most important of these was mounted in July 1967 with 'The World on Paper' as its theme. The emphasis was placed on early commercial mapmaking, the aim being to give an impression of the work of seventeenth-century Amsterdam publishers of maps, atlases, geographical books and nautical works. The authorities of the same museum collaborated in 1977 with the Art Gallery of Toronto in an exhibition entitled 'The Dutch Cityscape in the Seventeenth Century and its Sources'. Maps, town plans and views as well as paintings of the urban scene were featured and the exhibition was accompanied by a well-illustrated and informative catalogue with an interesting section on maps and profiles. A very specialized but fascinating exhibition arranged by the Museo Correr of Venice in 1982 displayed no fewer than 153 plans of the city ranging from an anonymous view of 1479 to an aerial photograph dated 1911. Such a display provided not only an insight into the development of Venice itself but also into the development of urban mapmaking. The exhibition was accompanied by a well-illustrated catalogue [221] with text and notes in Italian.

In the summer of 1984 the Map Room of the National Library of Scotland in Edinburgh mounted its first exhibition of maps. This was designed to show examples of the many different types of map available for consultation in the Map Room and together with a useful accompanying booklet written by the Superintendent of the Map Room, Margaret Wilkes, gave a revealing insight into the history, scope and services offered by the Map Room.

## Publishers' Catalogues

Map producers generally, whether official or commercial bodies, issue catalogues of the material they publish and it would be a sensible move for every curator of map collections or any systematic map collector to have himself placed on the various mailing lists. A particularly helpful list of catalogues received, together with the addresses of the issuing firms, appears in the Special Libraries Association, Geography and Maps Division *Bulletin* [375] from time to time. In addition to its general catalogue the Ordnance Survey of Britain issues a periodic *Map News & Review*, which gives details of any new map series and also announces forthcoming publications.

In the antiquarian map field dealers issue catalogues that are not only well produced but full of important information about the maps and atlases offered for sale. In this respect the catalogues of Robert Douwma Prints & Maps Ltd. of London are outstanding.

Prices of historical cartographical material can be established by constant reference to auction sales catalogues. Current prices vary considerably and in determining the right price for any item the following factors should be taken into consideration: (i) the significance of the particular map in the history of cartography; (ii) whether the item is in manuscript (in which case the price is considerably enhanced) or printed; (iii) the number of impressions originally printed; (iv) the number of surviving copies; (v) the printing process used: woodcut, copperplate engraving or lithography; (vi) the edition and state; and (vii) demand—this is partly governed by the foregoing factors but can also depend on the aesthetic value of the map and its condition or whether it is of particular local interest. When purchasing early maps generally it is wise to shop around well away from the area depicted on the map. Thus it would be better to look for maps of Cornwall or Devon in the north or Midlands of England and vice versa.

The Map Collector [327] offers a useful guide to the current values of early maps in two of its sections: 'Collectors' barometer' is compiled by Tessa Campbell and provides a continuous record of important items sold in auction at houses throughout the world, and 'Collectors' Marketplace' offers a forum in which collectors and dealers can advertise to a worldwide audience their sales items, wants or services offered. The December 1983 issue included an advertisement by a colourist of antique maps, a service not always very easy to locate.

## Topical Information Services for Mapmakers

For mapmakers and geographers it is important to have some relatively easy means of keeping up with topical events in order that their cartography may be as up to date as possible. Fortunately a regular topical map information service is available on an international scale from the Hungarian State Institute for Geodesy and Cartography. Production began in 1965 of a service called Cartactual [378] with Professor Sándor Rado as Editor-in-Chief. Cartactual is issued bimonthly and normally contains around sixteen pages of maps on varied topics: changes in administrative boundaries; changes in placename spelling; location of new industries, power-stations, dams, etc.; changes in communication systems; and similar data of this nature. It is available on subscription from Cartactual, P.O. Box 76, H-1367 Budapest, Hungary. The maps are particularly clear and easy to follow. Two colours are used, black for existing detail and a brick-red for changes.

The Geographical Digest is an annual publication which has been issued by Philip since 1964. It offers a service similar to that provided by Cartactual, albeit in a largely textual rather than cartographical format, and also includes useful population statistics.

## Copyright

The literature in cartographic journals and textbooks on the subject of copyright is very sparse. Such as does exist is often written from the copyright holder's point of view (e.g. Hunt, 1981; Nieweyer, 1976). Such reticence is due to two factors: first, copyright is a legal matter and as such tends to be avoided by non-specialists; and secondly, there is very little case-law directly affecting carto-graphic matters. Furthermore the situation varies from country to country, as do the laws of the different countries and thus, for example, U.S. government publications are not copyright—in direct contrast to the situation in the United Kingdom (Cerny, 1978), where the current legislation is the Copyright Act 1956, the Design and Copyright Act 1968, and the Copyright (Amendment) Act 1971.

The basis of this legislation is that original, published works are copyright for fifty years from the end of the calendar year in which the author dies, or in which the work is published. Copyright is vested in the author, or proprietor (in the case of a map published in a journal article), or employer (in the case of a map published by a firm), or the commissioner (in the case of a map commissioned from a cartographer) or the controller of HMSO (in the case of all material produced by the British government). Copyright is an item of property and can be bequeathed or sold to another person.

It is illegal to make copies of such copyright works without prior permission (and a copyright holder may charge a royalty and require an acknowledgement on the copy before granting such permission) in any shape or form and by any manner or means. Indeed, even in the case of a published map based on another—such as a street plan based on the large-scale Ordnance Survey plan— it is an infringement of the original copyright, in this case belonging to the Ordnance Survey (i.e. Crown copyright), as well as that of the publisher of the street plan, to make a copy without permission.

Since the majority of published maps of the United Kingdom are based on Ordnance Survey material and since the most frequent user requirement in a map collection is to copy part of a map, it follows that an awareness of the position of the Ordnance Survey on copyright is essential for all cartographers. That position is clearly set out in Ordnance Survey Information Leaflets 8, 23 and 45, which can be obtained from the Copyright Branch, Ordnance Survey. Except in specific details, such as the scale of royalties, these leaflets can also be used as a guide to the law in respect of copyright held by persons other than the Ordnance Survey.

Under the Copyright Act 1956 there are a few exceptions to the requirement to obtain permission before making a copy. These include the limited use of a part of a map for private study or research, reviews, criticism and reporting current events. The Ordnance Survey interprets this as 'up to four copies of an extract of up to 700 sq. cms. of mapping' (Ordnance Survey, 1975), and such an interpretation may hold good for other map publishers, although in the case of small-format maps, or those in atlases, the permitted area may be held to be smaller than 700 sq. cms. Map curators in most collections will find that many of the requests for copies of maps are in fact permitted by this exception, but it is important that the user is made aware of the law; an item of personal research may well be published at a future date, in which case the user will need to comply with the law's full requirements. Furthermore if the user is permitted to undertake the copying himself in an unsupervised situation, the institution providing the facility may be contravening the law. This feature of copyright is a largely untried area, and case-law is apparently inconsistent (*see* Phillips, 1980); and it may be necessary for all users to sign an indemnity to repay the institution any damages sustained in a legal action under the Copyright Act.

## Reproduction Fees

Whenever it is desired to use for publication purposes photographs of early maps or atlas sheets that are housed in a library, map collection or in private hands, permission must be sought from the owner, librarian or curator of the particular collection. To acquire permission is important even though the material may be much earlier than the fifty-year copyright period. While permission to reproduce is normally given, a reproduction fee may be charged and a stipulation made that the reproduction should include an acknowledgement made to specific requirements.

## References

Cerny, J. W. 'Awareness of maps as objects for copyright'. *American Cartographer*, **5**(1): 45–56 (1978).

Hunt, J. 'A brief guide to copyright, acknowledgements and priorities'. *SUC Bulletin*, **15**(1): 76–77 (1981).

Nieweyer, J. K. W. 'Auteursrecht en kartografie'. *Kartografisch Tijdschrift*, **2**(1): 16–18 (1976).

Ordnance Survey. *Copyright*. Ordnance Survey Leaflet New Series No. 8, para. 11 (1975).

Phillips, J. 'Authorising copyright infringements'. *Journal of Business Law*, pp. 109–117 (1980).

# 6   Map Care

## *Introduction*

For some disciplines a section in this book devoted to the circumstances and methods used in libraries to classify, catalogue and store the literature and to supply it to the user would be superfluous. Librarians use one of a very small number of classification schemes, and catalogues, whether card, microfiche or online, routinely provide access by author, title and subject as a minimum, while the more sophisticated systems provide additional multiple entries under a range of keywords, date of publication, language and so on.

Librarians are familiar with whatever system operates throughout their own library and can therefore advise the user even in subjects that are not their own specialization. Similarly the user, once he has familiarized himself with the basics of the system, should be able to use it competently.

Maps, however, are different. Most users are required to consult maps less frequently than books and have a consequent lesser facility with them. So the professional, in providing user services, may need to provide a considerable amount of interpretation for the user of what the map has to say and may also need to educate the user in those things the map cannot say. As the final section of this chapter will show, much of the literature on user services reflects this need to educate the users.

Map storage and conservation involves a number of problems not faced by the librarian, problems that are reflected in practices which affect the user also. These problems stem from the physical format of the maps, a format that precludes their storage on shelves, necessitates their transport in rolls or tubes, and frequently results in a higher proportion of maps being reserved for reference use rather than loans.

The classification and cataloguing of maps provides the greatest 'intellectual' difference between books and maps. One measure of this difference is seen in the necessity to provide a 258-page 'manual of interpretation' [650] to enable one of the standard cataloguing schemes to be used for maps. Whereas a librarian can expect to be asked for a book by a particular author or about a certain subject, the patron requesting a map is most likely to request a map covering a particular area. To the map user, the theme (or subject) is generally a secondary consideration and the author's name is generally irrelevant. Furthermore, since a large proportion of maps are stored in a closed-access area, the classification of maps does not serve its normal purpose of bringing works on a similar subject together on the shelf to aid browsing by a user. Indeed, owing to the physical constraints imposed by the maps themselves, some classification schemes use size or shape as the criterion for classification, a concept that is almost completely foreign to the librarian.

In map acquisition too, the professional curator faces difficulties that a librarian would find unusual. In contrast to the large number of private book publishing firms, the publishing of maps is concentrated into relatively few hands, many of which are organs of the state and which are frequently part of the military or security establishment. Consequently, maps are increasingly regarded as state secrets and are simply not available for some areas.

So maps are different and consequently, as the examples above have shown, they have to be treated differently. The jargon employed is different and the professionals are specialists. Indeed, except in North America, it is not common for them even to be called 'librarians'. 'Map Curator', 'Map Keeper', 'Map Officer', 'Cartothécaire' are all terms that might be encountered. In defence of the English language this chapter refers to 'Map Curator' and 'Map Care' rather than the illogical hybrids 'Map Librarian' and 'Map Librarianship'. These, like the similarly unfortunate term 'carto-bibliography', seem to have taken root, however, and are the normal usage in the United States of America and Canada. It is from North America, furthermore, that most of the writing on the subject emanates, supporting no fewer than four regular journals and a monograph series in addition to numerous texts on every aspect of the subject. Even in North America, however, the subject remains very largely in the purlieu of the academic institutions with the notable exception of the national collections.

However, the map curators of most American academic institutions have to have a public reference duty in addition to their internal roles. This duty comes about as a result of the fact that their institutions are publicly funded and contrasts with, say, the situation in Britain, where the university map curator's tasks are almost entirely internal.

## General Aspects of Map Care

It is against the background of difference from an assumed 'norm' that the literature of map care must be seen. Thus there is a tendency for textbooks to be aimed at 'those without specialist training' (Farrell and Desbarats, [605]) or 'the general librarian who has to deal with a map collection' (Nichols, [610]) or 'archivists without specialist knowledge of maps and drawings' (Ehrenberg, [604]). No advanced text is available and, since most of the authors are practising map curators with a full-time commitment to their institution, it may be that one is not to be expected. Indeed, it will be seen from the entries in Part III that the literature is not only pitched at an introductory level but is also very small in amount. The vast majority of the citations in the bibliographies relate to articles in journals rather than books, partly because, as already stated, the map curators do not have the time for fuller explanations of their subject. Another reason for the paucity of literature is that the subject as a whole is a young one. With the exceptions of a few collections of national importance there were virtually no map collections before the First World War and even by 1945 the numbers remained very small. The growth of the specialization can be followed in Ristow's book [612], a work which, although concentrating on North American practice, can be taken to represent the development elsewhere.

A third reason for the present balance of the literature follows partly from the previous reason. The specialization, being quite recent, is small both in its totality and in its individual institutions. Most map collections employ fewer than five persons and the 'one-man band' is common. Furthermore, because maps are different, pressure to integrate them within the library systems of the parent institutions has arisen only recently with the spread of the application of automated methods, principally in cataloguing. Consequently, many map collections can be described as idiosyncratic or experimental to a greater or lesser extent. Despite the increasing standardization of cataloguing rules it is likely that the relative independence of many map collections will ensure that this variety will be a feature of the specialism for many years to come, with a continuing stream of articles in the journals expounding and promoting different practices.

Current practice is not, however, the sole meat of the growing number of journals. The journals on map care *per se* are wholly the products of the various associations of map curators and thus, with the exception of the monographic Occasional Papers of the Western Association of Map Libraries, include news material, announcements and other items of a newsletter nature in addition to fuller papers. A very valuable part of this 'current awareness' side of the journals' work is the appearance in many of listings and reviews of recent publications: maps, atlases, books and papers. Books on related subjects such as geography, geology and exploration as well as cartography itself are covered by several of the journals, the *Bulletin* [624] of the Geography and Map Division of the Special Libraries Association being especially strong in this area. As its title suggests, this journal is addressed both to map curators and to geography librarians—a not uncommon combination of roles.

The content of the articles printed in the journals also reflects the interests of the societies' membership, so that in addition to articles concerned with practical issues there are a number which are of a bibliographic or carto-bibliographic nature. It was from such articles as these that the WAML's Occasional Paper series arose, providing carto-bibliographies of, for example, Fiji and Utah, or union listings of the Sanborn Fire Insurance Maps of North American towns and cities.

A further major interest of map curators lies in the history of cartography and the number of contributions on this topic reflects this interest. For this reason also many map collections take, or have access to, the journals and literature of the history of cartography and of antiquarian map collecting. In these too, as in the more general cartographic literature and that of librarianship, some items relevant to map curators are to be found.

## Acquisition

Logically map acquisition must be considered first, not merely for the obvious reason that without any acquisitions the care of maps cannot proceed, but also because the difficulties referred to in the introductory paragraphs need to be appreciated.

Patrons and even librarians brought up in most countries of Western Europe receive an exposure, both in school and in general experience, to maps—frequently of a high quality—which conditions them to expect a similar level of cartographic awareness and excellence worldwide. If Britain is taken as an example it is accepted as quite normal to be able to walk into a good stationer's or bookshop in any town and buy a 1:50,000 or 1:25,000 map of the local area published by the national survey (in Britain's case the Ordnance Survey) and for a selection of other areas. The Ordnance Survey additionally maintains a network of agents who provide a more comprehensive cover in areal terms, and for the local area in scale terms also. These agents can produce to order the full range of Ordnance Survey products.

Similar, or even greater, cartographic awareness and availability applies in many Western European countries. It is easy to understand, therefore, why patrons find it difficult to appreciate that for much of the world 1:250,000 or 1:200,000 is the best available scale and harder still for them to realize that even these scales are often unavailable for security reasons.

The reasoning by which an apparently growing number of nations restrict the availability of topographic information seems increasingly unrealistic given the growth of mapping from remote sensors, notably the American LANDSAT series and the French SPOT programme. Such restrictions continue to thwart the efforts of many map curators to serve the needs of their patrons.

Another difficulty facing map curators that will be unfamiliar to their counterparts in libraries is the complete lack of an equivalent to *British Books in Print*. Indeed, many of the bibliographic aids available to the librarian have no counterparts available to the map curator. It is symptomatic of this difficulty that the only real approximation to an international carto-bibliography is the catalogue of the West German map retailer Geo Center [629]. Not unnaturally this publication emphasizes the areas of greatest commercial interest to the company so that the Federal Republic of Germany, the Alps, and other European areas are provided with considerably more detail than the United States, for example. A further problem for map curators who try to use this catalogue as a carto-bibliography is that it gives only a minimum of publication information. This is an entirely reasonable means for the publisher to protect his own commercial interests but it does pinpoint the need for a genuine carto-bibliographical system. It can only be hoped that such a system will be developed from, for example, UNIMARC, the automated cataloguing system ([638], [640]).

In this climate of sparse or poor-quality information the listings of new maps published in many journals assume a particular significance. Especially useful lists are to be found in the *Bulletin* of the Western Association of Map Libraries and in the *Bulletin* [624] of the Geography and Map Division of the Special Libraries Association. In Britain a similar listing is provided in the *Cartographic Journal* [381]. This listing is compiled by the Curator of the Map Collection in the Bodleian Library, Oxford. The Bodleian is one of the copyright libraries and publishes its own acquisition list from which the list in the *Cartographic Journal* is a selection.

Lists of this kind are a valuable aid in acquisition, providing information not only on new mapping (their primary intention) but also on the areas of interest to different collections. This has a significant effect on acquisition policies, for many collections face increasingly stringent financial restraints on their acquisition policies. Inter-collection loans, co-operation on specialist areas, and exchange of surplus stocks all provide useful low-cost methods of maintaining the standards of service to map users.

More formal listings of the interests of different collections are to be found in the directories of map collections which are listed in Part III.

## Map Classification and Cataloguing

Probably more has been written on classification and cataloguing than on any other aspect of map care. The 'intellectual control' of a map collection, like that of a library or an archive, is an essential prerequisite to its full exploitation. The discussions have centred on how best to achieve this control and whether the control should be in a universally standardized form comparable to that for books and other items. Many of the works on classification and cataloguing are merely statements of the particular scheme employed in the author's institution or, more usually, developed by the author himself.

Although classification and cataloguing are traditionally mentioned in the same breath and, although a number of systems aim to combine the two operations, it is necessary to discuss them separately here for reasons of clarity.

### Cataloguing

For a librarian the exercise of cataloguing has the relatively simple aim of providing in a list the identity of all the material under his control. From such a list it can be ascertained for example whether a particular edition of a book is held. The catalogue thus provides the basis for checking stock, either as a routine exercise or prior to acquisition, without the necessity of physically examining the bookstock. Similarly the prospective reader can discover, or be informed, with a minimum of time and trouble whether the item he requires is held by the library. In a lending library the catalogue may also be linked to the loans system so that the reader may know not only whether the volume is held, but also whether it is in stock or on loan.

Numerous cataloguing systems have been developed. Among them *AACRII* (*Anglo-American cataloguing rules*, 2nd edition) is probably the most widely used. These *Rules* permit cataloguing at one of three 'levels' and Table 3 in the *Rules* shows which categories of information are required at each level. For most purposes Level 1 is adequate while Level 3 will be adopted only by the major national libraries and national bibliographies. The differences between books and maps are demonstrated most forcibly by the fact that not only is the chapter on 'cartographic materials' the longest in the book but, as mentioned above, a 258-page 'Manual of interpretation' [650] is necessary to make the rules usable in map collections. Clearly there are considerable difficulties in achieving *AACRII*'s aim of providing a standard form of catalogue for all classes of material.

One of the most far-reaching differences between books and maps is that a catalogue is not, for the majority of maps, the most efficient means of storing information about holdings. The reason is that the majority of maps are published as part of a map series. This concept is almost unknown in the book publishing world so a brief explanation is in order. A book, whether in one or more volumes, is considered as an entity, having an author or editor, a publisher and a date of publication. A number of subsequent editions may be published, containing a greater or lesser amount of revision, and in some cases these subsequent editions may have additional, or different, authors or editors, or even different publishers. The entity of the work, however, is never in question. By contrast the serial, exemplified by the journal of a learned society, is commonly published in, say, annual volumes each containing one or more parts or issues. Although retaining the same title ('*The Cartographic Journal*', for example), each volume can be regarded as a separate entity, having an editor and publisher who may or may not be the same as in previous or subsequent issues, and a date of publication. Such volumes are often provided with preliminaries and indices emphasizing their entity, an entity that is however circumscribed by a number of continuing features that can be covered by the term 'house style'.

The map series falls somewhere between these two concepts, differing in important respects from both. A series is conceived of as an entity, but an entity in which all the components themselves are separate entities. Although a series may have common specifications, therefore, these may be changed during the life of the series, or in particular components of the series. Furthermore it is most unusual for a map series to have an author in the accepted sense, while different components in the series will often have different publication histories.

Indeed it can be reasonably argued that catalogue entry is a most inefficient way of holding information about a map series. It is perhaps more logical to store data of a cartographic nature in a cartographic form using a visual index. It is certainly easier to answer the enquiry 'Have you got a map of ... ?' by using a visual index.

Thus, given that in most map collections the majority of the stock is formed of map series, it is likely that a catalogue will be used as a secondary, rather than a primary, tool by the curator—a situation that librarians would find most unusual.

## Classification

The predominance of the map series and particularly of the topographic map series is also important in classifying map collections. There is an even greater variety of classification schemes to be found in map collections than there is of cataloguing systems. Furthermore there is much less pressure towards uniformity, largely because the classification schemes are developed as a specific response to the peculiar needs of, and demands placed on, each collection. To take an extreme example: the classification scheme devised for a university geological map collection would be hardly likely to suit a collection associated with a County Record Office. The former could be based on the subject matter of the maps—geophysical, quaternary or tectonic, for example—or on the area portrayed—perhaps the Alps, the English Channel or Ireland. Since a County Record Office map collection is likely to consist almost wholly of topographic maps of the county or parts of it, classification by subject matter is unlikely to be helpful. Date or function—estate maps, tithe maps or electoral maps—are more appropriate bases for classification, or provenance. This last is the guiding principle of the intellectual control of a Records Office and would commonly provide the basis of a classification scheme. Many 'home-made' schemes have been developed, however. Some, where storage space is at a premium and most of the stock in single pieces, use size as the main criterion.

This kind of scheme, of storage by size, emphasizes the dual role of classification: not only does it provide a division of the stock into useful classes but these classes are frequently used to organize the storage area. The fullest expression of the way in which a classification scheme may be used to organize the storage area is to be found in the Parsons/GSGS 5307 scheme [646]. This was developed purely for internal use by Britain's War Office, now the Ministry of Defence. It was 'devised to provide an area-based classification which would be embodied in the library reference, which in turn would control filing to produce a strictly "geographical" arrangement; maps would be thereby housed in a uniform storage system which avoids the need for separate location references. The classification involves no subject element; some provision for retrieval by subject being made through the incorporation in the catalogue of a simple additional entry subject file.' Both derivatives of this scheme and comparable schemes are to be found throughout the world of map care.

Over the years the greatest difficulty in classifying maps has been concerned with multiple-class maps. While it is true that all maps can be classified by area, subject and date, it is also true that many maps can be classified into more than one area, subject or date. While librarians will be familiar with multiple subject work and extending this concept to the area portrayed is not difficult, the variety of dates available to the map curator is another important difference between books and maps. The variety of dates is particularly a problem among topographic maps that purport to show, within the limits of a chosen scale, the situation as it exists on the ground. In an area where development is occurring not only may the publication date be different from the survey or revision dates, but also there may be different survey and revision dates in varying parts of the map. Thus, maps can have a span of several years between the earliest date of survey or revision and the latest, and quite commonly have a further gap of two to five years before publication.

The multitude of possible headings for classification has made this activity a prime target for automation. Of particular value are systems such as Maplib [649], which can be interrogated by the user to search for the class of maps that fits his requirements. If he is fortunate this class will consist of one item!

## Storage and conservation

It is perhaps in the area of storage that the practices of the map curator and the librarian diverge most widely. The problems are completely different, as are the solutions, although examples do exist of maps that have been dissected and mounted so that they may be folded into a slip case which is then lettered on a false spine and stored 'book-like' on a shelf. With the minor exceptions noted below, the map curator's aim is to store all maps flat with as few folds as possible, a procedure that has two advantages. First, it is a conservation measure since the more folding and unfolding a sheet of paper receives, the weaker it becomes. Secondly, it increases the utility of the map for making measurements since folds cause a change in the size of the sheet paper. Most map collections also include a small number of folded copies of maps that are easily replaced and usually used as loan copies for use 'in the field'. It is, however, the storage of flat sheets that provides the subject matter of the bulk of the literature. Although this is not extensive it is largely composed of articles extolling the virtue of vertical, as opposed to horizontal storage. In fact, for most purposes the latter is to be preferred, particularly if there are restrictions on the space available for storage purposes. Small numbers of sheets are stored together in acid-free paper folders, which cut down the dust and reduce the amount of handling of individual sheets.

Such folders can be used in some of the vertical storage systems but too many of these are suspension systems that rely on either the application of self-adhesive tape to the maps or the punching of holes for the suspension bars. Such systems clearly damage the maps and probably also affect their dimensional stability. These defects, however, do not seem to dampen the ardour of their proponents.

It will be appreciated that most storage methods are adopted, at least in part, for their contribution to the conservation of the individual items for, as in all conservation methods, prevention is better than cure. The presence in many map collections of antiquarian or delicate 'middle-aged' mapping is a constant reminder to all map curators of the need for conservation. It is likely, however, that only a few map collections have direct access to conservation services of a sufficient capability. Most curators have to concentrate their resources on the point of greatest need and attempt to control the damage to other items by reducing their availability to users.

An important feature of the literature is the discussion of the appropriate air conditions for storage. The lead is frequently taken by the archival collections, but although the knowledge of the recommended practices is widely disseminated, a very large proportion of collections are not housed in ideal conditions. Conservation, for most map curators, has to be a matter of compromise between the conservation of the maps and the provision of a service to the users of the map collection.

## Reader services

The provision of reader services is largely neglected by the literature. This neglect must be due partly to the diverse nature of the services that map curators, particularly those having only one or two members of staff (which means all but the largest institutions), are called upon to provide. Amongst these are giving answers to questions such as the following:

1. Where is ...?
2. How do I get to ...?
3. Have you got a map of ...?
4. What maps are there of ...?
5. How do I use this piece of equipment?
6. What does this symbol mean?
7. What is the English for this word?

Provision of the answers can clearly vary from a rapid search in a gazetteer or dictionary to courses in map reading or map use. A map curator who has other responsibilities such as air photos or periodical and offprint collections needs to be even more versatile. Even among those curators whose charge is solely one of maps, the range of experience needs to encompass not only the use and interpretation of topographic maps but also of geological maps, geomorphological maps, distribution maps of various types and indeed of every type of map in the collection. A working knowledge of all the subjects covered by the maps, or at least of the ways in which maps may help the study of these subjects, is expected of the map curator by the users of his collection. Where this collection forms part of a teaching establishment, questions on cartographic techniques and aesthetics are often added to those of interpretation.

In the face of expectations such as these it may well be asked whether it is possible to begin to offer reader services. Whether it is depends on the professional attitude of the curator and his own interest in his chosen speciality. The primary requirement is an exhaustive knowledge of the collection. Armed with such knowledge the curator is immediately aware as to whether a specific enquiry can be answered from his own collection or whether the enquirer must be advised to try elsewhere. Whether or not this knowledge is directly held or has to be extracted from a catalogue or index is not of great importance. What does matter is the speed and accuracy with which the information is provided, for it is on these aspects of the service that users will judge the collection and its curator rather than on any technicalities of the classification system.

This matter seems largely to be taken for granted in the existing literature, which is unfortunate as the quality of the service offered is an important aspect of the public relations of the map collection. It is important to attract and retain users, particularly in organizations that link user-figures to budgets. Even in those institutions which do not make a direct link between the two it is common for map curators to keep records of user enquiries on a regular basis. If sufficiently detailed these can provide information, not only on the numbers and sources of users, but also on their needs and whether or not the collection was able to satisfy them. This information in turn provides a valuable check on the validity of the collection's acquisition policy, which may need to be amended from time to time in the light of changing user requirements.

PART II

Annotated Bibliography of Reference
Sources: History of Cartography

# Information Sources on the History of Cartography

## General Sources

### Bibliographies

1 *The bibliography of cartography.* Washington, D.C.: The Library of Congress, Geography and Map Division, 1973. 5 vols. (Distributed by G. K. Hall, 70 Lincoln Street, Boston, Massachusetts 02111.)
A unique analytic index to the literature of cartography. Made from approximately 106,000 cards originated in the 1870s by Philip Lee Phillips. Especially rich for the period 1895–1922. An invaluable reference work for libraries, government agencies, universities and other establishments involving geography, cartography, geology and earth sciences.

2 *National Ocean Survey cartobibliography: Civil War collection.* Compiled by Map Library, Physical Science Services Branch, Scientific Services Division, Rockville, Maryland: National Ocean Survey, 1980. 64pp. (NOAA–NOS Technical Service Publication, available from the Superintendent of Documents, U.S. Government Printing Service, Washington, D.C. 20402.)
Relates to the part of the National Survey map collection which consists of 8,500 historical maps. It is organized by state, region, theme, atlas or series and provides an important reference work for any map library.

3 **Ristow, Walter W.** (compiler). *Guide to the history of cartography: an annotated list of references on the history of maps and mapmaking.* 3rd edn. Washington, D.C.: Library of Congress, 1973. 96pp.
The first and second editions were issued under the title *A guide to historical cartography.* The present edition includes 398 references in alphabetical order by author or by first word in the title with Library of Congress call numbers for all

entries. Annotations amplify the bibliographic descriptions and draw attention to the special significance of particular references. Has an alphabetical index. A handy and compact reference work for map librarians, historians of cartography and cartographers.

**4 Koeman, Cornelis.** *Atlantes Neerlandici. Bibliography of terrestrial, maritime and celestial atlases and pilot books, published in the Netherlands up to 1880.* Amsterdam: Theatrum Orbis Terrarum, 1967–71. 5 vols.
An important catalogue of maps and atlases published in the Netherlands between 1570 and 1880. Describes the maps by title, imprint, size and scale. Indicates major revisions of the plates. Valuable for all students of the history of mapmaking and for librarians of collections in which early atlases form a significant part.

**5 Mickwitz, Ann-Mari** and **Leena Miekkavaara.** *The A. E. Nordenskiöld Collection in the Helsinki University Library. Annotated catalogue of maps made up to 1800*; Vol. 1, *Atlases A–J*; Vol. 2, *Atlases K–Z*. 1981. Vol. 3, *Books containing maps, loose sheets, and addenda to vols. 1 and 2.* 1984. Helsinki: University Library. (Distributed by Almquist & Wiksell International, Stockholm and Humanities Press, Atlantic Highlands, New Jersey)
Bibliography of the collection assembled by the explorer and cartographic historian, A. E. Nordenskiöld. Includes an introduction to his life and works. Lists 119 atlases under author's name with full descriptions.

**6 National Maritime Museum.** *Atlases and cartography.* London: HMSO, 1971. Vol. 3 of the Catalogue of the Library of the National Maritime Museum, Greenwich.
With an introduction by the librarian, Michael Sanderson. Individual mapmakers are listed alphabetically within national groups.

### Guide to bibliographies

**7 Drazniowsky, Roman.** 'Bibliographies as tools for map acquisition and map compilation'. *The Cartographer*, **3**(2): 138–144 (December 1966).
The map curator of the American Geographical Society examines a variety of information sources within the framework of their relevance to: (a) maps and atlases, globes; (b) general cartography including history of cartography; (c) bibliographical aids and gazetteers. With a selective bibliography of fifty-two entries. All map users, but particularly librarians and map curators, will find this a useful paper.

### Major catalogues and inventories

**8 British Museum, Department of Printed Books.** *Catalogue of printed maps, charts and plans. Photolithographic edition complete to 1964.* London: British Museum, 1967. 15 vols.
Corrections and additions, 1968. 55pp.
One of the most important reference tools for historians, cartographic historians,

map curators, librarians, bibliographers, geographers and all whose work involves the use of early maps. Covers maps, atlases, globes and other material preserved in the Map Room as well as cartographic material preserved in the Department of Printed Books and the Department of Oriental Printed Books and Manuscripts. Includes well over 200,000 entries.

**9 British Museum, Department of Manuscripts.** *Catalogue of the manuscript maps, charts and plans, and of the topographical drawings in the British Museum.* London. Printed by order of the Trustees. 1844–61. 3 vols.
Describes the cartographic material in the Department of Manuscripts, those attached to the library of King George III, and the Print Room collections.

**10 Public Record Office.** *Maps and plans in the Public Record Office.* Vol. 1, *British Isles,* c. *1410–1860.* Vol. 2, *America and West Indies.* Vol. 3, *Africa.* London: HMSO. Vol.1, 1967; Vol. 2, 1974; Vol. 3, 1982.
Three volumes that should be essential for all major libraries.

**11 U.S. Library of Congress.** *A list of geographical atlases in the Library of Congress, with bibliographical notes.* Washington, D.C.: Government Printing Office. Vols. 1–4 by Philip Lee Phillips. Vols. 5–8 by Clara Egli LeGear. 1909–74. 8 vols. Indexes.
Lists over 18,000 atlases with bibliographical information and analytical notes. The most comprehensive work on historical atlases. A reprint of vols. 1–4 was made by Theatrum Orbis Terrarum, Amsterdam in 1970.

## Major works

**12 Bagrow, Leo.** *Die Geschichte der Kartographie.* Berlin: Safari Verlag, 1951. 383pp.

**13 Bagrow, Leo.** *History of cartography.* Revised and enlarged by R. A. Skelton. London: C. A. Watts, 1964. Cambridge, Massachusetts: Harvard University Press, 1964.
An English translation of the previous, German work. Probably still the best introduction to the subject. Very well illustrated in colour and monochrome. Bibliographical notes. Useful for historians, cartographic historians, geographers, librarians and all map users.

**14 Blakemore, M. J.** and **J. B. Harley.** *Concepts in the history of cartography: a review and perspective.* Edited by Edward H. Dahl. *Cartographica,* **17**(4). Monograph 26. Toronto: University of Toronto Press, 1980. 120pp.
An important review of trends in writing on cartographic history, the central theme being the lack of philosophical and methodological writing on early maps. Includes an important list of references with over 300 citations. Essential for carto-bibliographers and historians. Volume 19 No. 1 (pp. 66–96) of the same periodical carries six reviews of Blakemore and Harley's Monograph, together with their reply.

**15 Bricker, Charles.** *Landmarks of mapmaking; an illustrated survey of maps and* *mapmakers. Maps chosen and displayed by R. V. Tooley; text written by Charles Bricker; preface by Gerald Roe Crone.* Brussels: Elsevier-Sequoia, 1968. 220pp.

**16 Bricker, Charles.** *A history of cartography: 2,500 years of maps and mapmakers.* London: Thames & Hudson, 1969. 220pp.

The contents of this and the above book by the same author are identical, only the title being changed from the Elsevier-Sequoia edition (also published in London by Phaidon). The illustrations are superb, with several full-scale folding colour facsimile maps and many illustrations in monochrome. For laymen as well as cartographic historians and map collectors.

**17 Brown, Lloyd A.** *The story of maps.* Boston: Little, Brown, 1949. 397pp. Reprinted by Dover, New York. 1979.

One of the best introductions to the history of mapmaking. Includes a useful bibliography.

**18 Cortesão, Armando.** *History of Portuguese cartography.* Lisbon: Junta de Investigações de Ultramar. Vol. 1, 1969; vol. 2, 1971.

In vol. 1 Cortesão looks at the period up to the fourteenth century and in vol. 2 at the fourteenth and fifteenth centuries. The second volume includes two chapters on the history of astronomical navigation. There is a bibliography on pp. 307–312 of vol. 1. Of prime interest to historians and cartographic historians but useful also to geographers and hydrographers, students of navigation.

**19 Cortesão, Armando.** *Cartografia e cartógrafos portugueses dos séculos XV e XVI.* Lisbon: Edição da 'Seara Nova', 1935. 2 vols.

A study of Portuguese nautical charts and their makers which is one of the most important works on early cartography. Includes bibliographical references and extensive index as well as facsimile reproductions of 46 charts.

**20 Cortesão, Armando** and **Avelino Teixeira da Mota.** *Portugaliae monumenta cartographica.* Lisbon: Portuguese government, 1960–62. 6 vols.

One of the finest works on the history of cartographic development. Includes splendid colour reproductions. Text in English and Portuguese. Volume 1 looks at cartography to the mid sixteenth century; vol. 2 at the mid sixteenth century to its last decade; vol. 3 at the end of the sixteenth century; vols. 4 and 5 at the seventeenth century; vol. 6 contains the comprehensive index.

**21 Crone, Gerald Roe.** *Maps and their makers; an introduction to the history of cartography.* 4th revised edn. London: Hutchinson, 1968. 181pp.
Revised edition, larger format. Folkestone: Wm. Dawson & Sons, 1978; Shoe String, Hamden, Connecticut: Archon Books, 1978. 152 pp.

Originally published in 1953, this scholarly work by the former head of the Royal Geographical Society's map room looks at maps as scientific reports, as historical documents, as research tools for historians and geographers, and as objects of art.

**22 Fordham,** *Sir* **Herbert George.** *Maps, their history, characteristics and uses.* 2nd edn. Cambridge: Cambridge University Press, 1927. 83pp.

The text of a series of five lectures to Cambridge teachers with the object of stimulating their interest in cartography.

**23 Fordham,** *Sir* **Herbert George.** *Studies in carto-bibliography, British and French, and in the bibliography of itineraries and road-books.* Oxford: Clarendon Press, 1914. Reprint, London: Dawsons of Pall Mall, 1969. 180pp.
A collection of papers from various sources in which Fordham attempts to create interest in bibliographical matters connected with geographical and historical study.

**24 Koeman, Cornelis.** *Collections of maps and atlases in the Netherlands: their history and present state.* Leiden: E. J. Brill, 1961. 301pp. (*Imago Mundi* Supplement III.)
Scientific study in the history of cartography has been hindered by our limited knowledge of the source material. Professor Koeman has tried to remedy this by publishing a list of map collections in the Netherlands; by formulating a schematic outline of the bibliography of atlases published in the Low Countries before 1800; and by studying the character of map collecting through the centuries. Important for historians of cartography, map collectors, geographers and all users of early maps and atlases.

**25 Nordenskiöld, Nils Adolf Erik.** *Facsimile-atlas to the early history of cartography with reproductions of the most important maps printed in the XV and XVI centuries.* Originally published in Stockholm, 1889. Reprint, New York: Dover Publications. 1973. 142pp. plus 52 plates and 84 illustrations in text.
A highly important and useful work which was the first systematic approach to the study of cartographic history. Reproduces 169 of the most important maps printed before 1600, including all from the 1490 edition of Ptolemy. Particularly useful for students concerned with the geography and history of the Age of Discovery.

**26 Skelton, Raleigh A.** *Decorative printed maps of the 15th to 18th centuries: a revised edition of Old Decorative Maps and Charts, by A. L. Humphreys. With eighty four reproductions and a new text by R. A. Skelton.* London: Spring Books, 1966. 80pp.
A well-illustrated introduction with good bibliographical material. Of interest to map collectors and laymen as well as historians, carto-historians and geographers.

**27 Taylor, Eva Germaine Rimington.** *Late Tudor and early Stuart geography, 1583–1650.* London: Methuen, 1934. 332pp. Reprint, New York: Octagon Books, 1968.
Includes a very extensive bibliography. Important work for historical geographers and carto-historians.

**28 Taylor, Eva Germaine Rimington.** *Tudor geography, 1485–1583.* London: Methuen, 1930. 290pp. Reprint, New York: Octagon Books, 1968.
Contains much information about cartography and survey during a period of considerable progress.

**29 Wallis, H.** and **A. H. Robinson** (eds.). *Cartographical innovations: an*

*international handbook of mapping terms to 1900.* Tring, Hertfordshire: Map Collector Publications. (In the press.)
An important history of cartographical innovations and their diffusion, with definitions and bibliographies for each term.

**30 Wieder, Frederik Caspar** (ed.). *Monumenta cartographica: reproductions of unique and rare maps, plans and views in the actual size of the originals; accompanied by cartographical monographs.* The Hague: Martinus Nijhoff, 1925–33. 5 vols.
This lavish work is particularly important for the light it throws on Dutch contributions to cartography and exploration. A valuable reference for carto-historians, historians, geographers, collectors.

**31 Wilford, John Noble.** *The mapmakers.* Junction Books: London, 1981. 414pp.
Examines maps as a means of communication, and chronicles the exploits of mapmakers right through to the present day. Especially good on American mapping and on contemporary technology. A very readable account for geographers, historians, cartographers and surveyors.

## Research directory

**32 Clark, P. K., Elia M. J. Campbell** and **E. A. Clutton** (eds.). *International directory of current research in the history of cartography and carto-bibliography.* Available from Geo Books, 34 Duke Street, Norwich NR3 3AP, England.

## History of cartography as a field of study

**33 Woodward, David.** 'The study of the history of cartography: a suggested framework'. *The American Cartographer*, **1**(2): 101–115 (1974).
Introduces a matrix in which the columns distinguish between the history of cartographic production with its elements of personnel, techniques and tools, and the history of cartographic products. The rows show the various stages of the cartographic process: information gathering, processing, document distribution, document use. Woodward also clarifies important terms and phrases and provides a simple definition of the field as a basis for discussion. Basically for cartographic historians.

## Journal

**34** *Imago Mundi.* c/o King's College, London. 1935–.
A valuable source of reference material on early cartography. Contains scholarly, well-illustrated articles, reviews and news items. Annual issues with irregular supplementary issues on specific topics. Essential for historians of cartography and of great interest to collectors.

## Visual aids

**35 Kish, George.** *History of cartography.* New York: Harper & Row, 1973. 220 colour slides 2 × 2 inches with 49-page booklet.
Well-reproduced set of colour slides which are classified under twenty-two headings. Illustrations of astronomical and surveying instruments are included as well as early maps. The accompanying booklet provides brief notes about each slide, a table of contents and an index. The collection is a valuable teaching aid

for courses in the history of cartography, discovery and exploration, geography and has some significance in the teaching of graphic art and design.

**36 Kish, George.** *The discovery and settlement of North America, 1500–1865.* New York: Harper & Row. 200 colour slides, 35 mm.
The slides are accompanied by an explanatory text and the set makes a valuable aid in teaching courses on North American history.

**37 British Library.** *Geographical knowledge as reflected in cartography.* London: The British Library.
A collection of eighty colour slides, derived from various sources, which would be helpful in geography teaching at advanced and university level.

**38 Bodleian Library**, Oxford. Film strips on cartographic topics:
*Cartographic treasures Exhibition I.* (37 frames)
*Cartographic treasures Exhibition II, British regional mapping* (35 frames)
*Early world maps, 11th–15th century* (12 frames)
*Portalano, Venice, c. 1400* (20 frames)
*The Gough Map* (10 frames)
*Large scale details of the Gough Map, from Central England* (12 frames)
*Gough Map details, North Britain* (6 frames)
*Atlas composé de mappamondes, de portulans et de cartes hydrographiques et historiques. Recueillies et graves sous la direction du Victomte de Santarem* (17 frames)
*Maps from Corpus Christi College, Oxford* (24 frames)
*Maps of Wales. 14th to 15th centuries* (5 frames)
*Portulan charts, 1660–1741.* Selected by Professor T. R. Smith (15 frames)
*Oxfordshire maps, 1605–1941.* (16 frames)
*Buckinghamshire maps: from Saxton to early Ordnance Survey* (15 frames)
*Maps from 7 MSS. of 13th to 18th centuries* (13 frames)
*Portuguese portolanos c. 1550 and a Carta Atlantica c. 1570* (15 frames)
The above form extremely useful teaching material for courses on the history of cartography or discovery and exploration, etc.

**39 National Map Collection of Canada**, Ottawa. Series of 10-slide sets of historic maps of Canada.

**40** The **Hermon Dunlap Smith Center for the History of Cartography** at the Newberry Library, Chicago has produced sets of 6 colour slides from atlases by each of the following important mapmakers: Ptolemy, 1482; Braun and Hogenberg, 1574; Saxton, 1577; Ortelius, 1606; Mercator, 1611. Available from the Bookshop, Newberry Library, 60 West Walton Street, Chicago, Illinois 60610.

# Individual Mapmakers

### Dictionaries
**41 Bonacker, W.** *Kartenmacher aller Länder und Zeiten.* Stuttgart: H. Hiersemann, 1966. 243pp.

Bonacker lists 6,350 mapmakers including all persons engaged in the production, publication and distribution of maps and charts, i.e. engravers, lithographers, printers and map-sellers; also collectors, map historians and bibliographers. There is an introductory essay in German and English.

**42 Eden, Peter** (ed.). *Dictionary of land surveyors and local cartographers of Great Britain and Ireland, 1550–1850*. Compiled from a variety of sources by Francis Steer (and others) and edited by Peter Eden. Folkestone: Wm. Dawson & Sons, 1975–76. 377pp. plus Supplement, 1979, pp. 378–528.
Contains around 10,000 entries relating to persons likely to have measured land or made maps of land in areas of less than a complete county in Great Britain or Ireland over three centuries.

**43 Tooley, Ronald Vere.** *Tooley's dictionary of mapmakers*. Tring, Hertfordshire: Map Collector Publications; New York: Alan R. Liss; Amsterdam: Meridian, 1979. 684pp.
The most comprehensive dictionary of individuals concerned with map production and an important reference work for historians, carto-historians, map collectors, geographers, and others. Contains over 20,000 entries including cartographers, geographers, engravers, publishers and others. The entries provide just enough information: name, date of birth and death when known; addresses; list of selected works. Preface by Dr. Helen Wallis of the Map Room, British Library.

### Selected works

**44 Averdunk, Heinrich** and **J. Müller-Reinard.** *Gerhard Mercator und die Geographen unter seinen Nachkommen*. Gotha: Justus Perthes, 1914. 198pp.
An important study in German of the work of the great Flemish geographer and his contemporaries.

**45 Bagrow, Leo.** *A. Ortelii Catalogus cartographorum*. Gotha: Justus Perthes, 1928–30. 2 vols.
A study of Ortelius, contemporary of Mercator, and his atlas *Theatrum Orbis Terrarum* including information on those mapmakers whose work is included in *Theatrum*.

**46 Evans, Geraint N. D.** *North American soldier, hydrographer, governor; the public careers of J. F. W. Des Barres, 1721–1824*. New Haven, Connecticut: Yale University, 1965. 305pp.
A biography of Des Barres with emphasis on the superb charts of eastern North America reproduced in *Atlantic Neptune*, the work for which Des Barres is best known.

**47 Evans, Ifor M.** and **Heather Lawrence.** *Christopher Saxton: Elizabethan mapmaker*. Wakefield: Wakefield Historical Publications; London: The Holland Press, 1979. 186pp.
A major work on Saxton, the first man to survey and map the counties of

England and Wales. Excellent bibliography. Essential reading for carto-historians, geographers, map collectors and map curators.

**48 Fordham, Herbert George.** *John Cary: engraver, map, chart and print-seller and globe-maker 1754 to 1835.* First published Cambridge: Cambridge University Press, 1925. Reprint, Folkestone: Wm. Dawson & Sons, 1976. 139pp.
Important study of the leading British mapmaker and engraver of the late eighteenth and early nineteenth centuries. With a detailed catalogue of the work of Cary and his successors.

**49 Guthorn, Peter J.** *American maps and mapmakers of the Revolution.* Monmouth Beach, New Jersey: Philip Freneau Press, 1966. 48pp.
Contains biographies of approximately fifty mapmakers connected with the Revolutionary armies.

**50 Knauer, Elfriede Regina.** *Die Carta Marina des Olaus Magnus von 1539.* Göttingen: Gratia-Verlag, 1981. 151pp.
A study in German of the Swedish cartographer Olaus Magnus and his important nine-sheet woodcut map.

**51 Koeman, Cornelis.** *The history of Abraham Ortelius and 'Theatrum Orbis Terrarum'.* New York: American Elsevier Publishing Co., 1964. 64 pp.
Originally published as an introduction to a facsimile of *Theatrum*.

**52 Koeman, Cornelis.** *The history of Lucas Janszoon Waghenaer and his 'Spieghel der zeevaerdt'.* New York: American Elsevier Publishing Co., 1964. 72pp.
Originally published as an introduction to a facsimile of *Spieghel der zeevaerdt* [Mariners' mirror].

**53 Koeman, Cornelis.** *Joan Blaeu and his grand atlas.* London: George Philip & Son; Amsterdam: Theatrum Orbis Terrarum, 1970. 114pp.
A short, well-illustrated biography of the Blaeu family and their contribution to cartographic development.

**54 Koeman, Cornelis.** *The sea on paper: the story of the van Keulens and their 'sea-torch'.* Amsterdam: Theatrum Orbis Terrarum, 1972. 92pp.
Includes material on Dutch pilot books since 1584 as well as a biography of the van Keulens.

**55 Ortroy, Fernand Gratien van.** *Bibliographie de l'oeuvre de Pierre Apian.* Originally published in Besançon by P. Jacquin, 1902. Reprint, Amsterdam: Meridian, 1963. 120pp.
An important bibliography of the works of Peter Apian together with a biography.

**56 Ortroy, Fernand Gratien van.** *Bibliographie sommaire de l'oeuvre mercatorienne.* Paris: Champion, 1918–20. 80pp.
A bibliography of works by the Mercator family.

**57  Osley, A. S.** *Mercator: a monograph on the lettering of maps etc. in the 16th century Netherlands.* London: Faber & Faber, 1969. 209pp.
Biographical material on Mercator plus an examination of his influence not only on the lettering of maps and globes but on book production, scientific instruments and engraving. A finely illustrated volume of interest to cartographers, carto-historians, calligraphers, typographers and designers.

**58 Sandler, Christian.** *Johann Baptista Homann, Matthäus Seutter und ihre Landkarten; ein Beitrag zur Geschichte der Kartographie.* Amsterdam: Meridian, 97pp.
Biographies of the two most significant eighteenth-century German mapmakers and information concerning their publications.

**59  Sanz López, Carlos.** *La 'Geographia' de Ptolomeo, ampliada con los primeros mapas impresos de América, desde 1507; estudio bibliográfico y crítico, con el catálogo de las ediciónes aparecidas desde 1475 a 1883, comentado e ilustrado.* Madrid: Librería General V. Suarez, 1959. 281pp.
An important discussion of Ptolemy's *Geographia* and its many editions by a Spanish scholar and cartographic historian.

**60  Stahl, William H.** *Ptolemy's 'Geography': a select bibliography.* New York: The New York Public Library, 1953. 86pp.
A useful bibliography of literature relating to Ptolemy and his work—1,464 references in all.

**61  Stevens, Henry N.** *Ptolemy's 'Geography': a brief acccount of all the printed editions down to 1730.* Amsterdam: Theatrum Orbis Terrarum, 1972. 62pp. (a reprint of the original publication, London: Henry Stevens, Son & Stiles, 1908).
Stevens lists and discusses the various editions of *Geographia* preserved in the Edward E. Ayer Collection, Newberry Library, Chicago.

**62 Tyacke, Sarah** and **John Huddy.** *Christopher Saxton and Tudor mapmaking.* London: The British Library, 1980. 64pp.
A well-illustrated volume written to accompany the exhibition of the same name. Includes useful information about contemporary surveying practices.

**63  Van Eerde, Katherine S.** *John Ogilby and the taste of his times.* Folkestone: Wm. Dawson & Sons, 1976. 183pp.
An important study of the life and work of the cartographer John Ogilby, famous for his pioneering road-book, *Britannia* (1675), in which he used the one inch to one mile scale for the first time. Includes a bibliography of Ogilby's works.

## The Map Trade

**64 Tyacke, Sarah.** *London map-sellers 1660–1720.* Tring, Hertfordshire: Map Collector Publications, 1978. 159pp.

Based on map-sellers' advertisements in *The London Gazette*, 1668–1719. Has biographical notes on individual map-sellers with maps to show the location of their premises. An important reference work for historians, carto-historians, sociologists, urban scholars and geographers. Contains useful lists of references.

**65 Tyacke, Sarah.** 'Map-sellers and the London map trade *c*. 1650–1710'. In H. Wallis and S. Tyacke (eds), *My head is a map*. London: Francis Edwards and Carter Press, 1973. 148pp.

**66 Pedley, M. S.** 'The map trade in Paris, 1650–1825'. *Imago Mundi*, **33**: 33–45 (1981).
This paper looks at the location of cartographers, engravers and mapsellers during a period when French cartography was at its best. Pedley emphasizes the close connections between cartographers, engravers and booksellers. With excellent maps of locations. The research is taken further in a subsequent article, 'New light on an old atlas'. *Imago Mundi*, **36**: 48–63 (1984).

### Trade catalogue

**67** *Sayer & Bennett's Catalogue of prints for 1775*. Reprinted, London: Holland Press, 1970.
An interesting facsimile reproduction of that rare phenomenon, an eighteenth-century trade catalogue of prints. Has a section entitled 'Useful and correct maps and charts' and includes prices.

# Discovery and Exploration

### Explorers' maps

**68 Skelton, R. A.** *Explorers' maps: chapters in the cartographic record of geographical discovery.* London, New York, Sydney, Toronto: Spring Books, 1970. 337pp. (Reprint of an original edition, London: Routledge & Kegan Paul, 1958.)
Traces the story of European exploration through the maps and charts of medieval travellers and later through the accurate maps and charts of eighteenth- and nineteenth-century surveyors. Designed to interest the general reader as well as historians, geographers and cartography students.

### Selected general works

**69 Albion, Robert G.** (ed.). *Exploration and discovery.* New York: Macmillan, 1965. 151pp.
Consists of thirteen papers with bibliographic references for each chapter.

**70 Albuquerque, Luis G. M. de.** *Introduçao a historia dos descrobrimentos.* Coimbra: Atlantida, 1962. 429pp.
A study in Portuguese with emphasis on Portuguese discoveries. Bibliography on pp. 401–426.

**71 Beazley, Charles R.** *The dawn of modern geography . . . a history of exploration and*

*geographical science ... with reproductions of the principal maps of the time.* London: John Murray, 1905–06. 3 vols.
A scholarly work on medieval geography with emphasis on discovery. Contains many illustrations of early maps with discussions on the significance of each.

**72 Debenham, Frank.** *Discovery and exploration: an atlas–history of man's journeys into the unknown.* London: Paul Hamlyn, 1960. 272pp. (Originally published, Stuttgart: Chr. Belser, 1960.)
A finely illustrated history of discovery and exploration aimed at the general reader. Includes reproductions of contemporary maps and prints to show the state of world knowledge at different periods. Maps show the successive stages of penetration of each continent. Route maps show the journeys of individual explorers.

**73 Heawood, Edward.** *A history of geographical discovery in the seventeenth and eighteenth centuries.* New York: Octagon Books, 1965. 475pp.
A reprint of an earlier edition published by Cambridge University Press in 1912.

**74 Hennig, Richard** (ed.). *Terrae incognitae; eine Zusammenstellung und kritische Bewertung der wichtigsten vorkolumbischen Entdeckungsreisen an Hand der darüber vorliegenden Originalberichte.* Leiden: E. J. Brill, 1944–56. 4 vols.
The first volume covers the ancient world to Ptolemy; second volume the period AD 200–1200; third volume 1200–1415; fourth volume 1416–97. A valuable reference work for historians, geographers and students of cartographic history.

**75 Penrose, Boies.** *Travel and discovery in the Renaissance. 1420–1620.* Cambridge, Massachusetts: Harvard University Press, 1955. 376pp.
Includes a good twenty-page bibliography and a chapter on the cartography of the Renaissance.

**76 Sharaf, 'Abd al-'Aziz Turayh.** *A short history of geographical discovery.* Alexandria: M. Zaki el Mahdy, 1963. 417pp.
Emphasizes the Muslim contribution to discovery and exploration. Includes a useful bibliography on pp. 378–395.


*Periodicals*

**77** The volumes of the **Hakluyt Society**, established in 1846, feature original narratives of voyages, travels, expeditions. Membership is open to all who are interested in the literature of travel and the history of geographical science and discovery. Two volumes are distributed to members each year. Information from the Administrative Assistant, Hakluyt Society, The British Library, Great Russell Street, London WC1B 3DG, England.

**78** *Terrae Incognitae, the annals of the Society for the History of Discoveries.* Amsterdam: N. Israel.
Concerned with exploration, techniques and literature of discoveries.

# The Developing World Map

**79 Alavi, S. M. Ziauddin.** *Arab geography in the ninth and tenth centuries.* Aligarh: Aligarh Muslim University, 1965. 134pp.
Reproductions of several early Arab world maps are included, together with some discussion of Arab cartography.

**80 Almagià, Roberto** and **Marcel Destombes.** *Mappemondes AD 1200–1500. Catalogue préparé par la Commission des Cartes Anciennes de l'Union Géographique Internationale.* Amsterdam: N. Israel, 1964. 280pp.
The most important catalogue of medieval *mappae mundi* selected by a commission of experts on the history of cartography for the International Geographical Union. Includes forty pages of plates and an index showing the location of the manuscripts cited. Also a valuable general bibliography of approximately 200 items.

**81 Andrews, Michael C.** 'The study and classification of medieval *mappae mundi*'. *Archaeologia*, **75**: 61–76 (1926).
A useful introduction to medieval world maps which the author classifies into families, divisions, genera and species.

**82 Kimble, George H. T.** *Geography in the Middle Ages.* New York: Russell & Russell, 1968. 272pp. (Originally published, London: Methuen, 1938.)
Includes useful information on medieval world maps.

**83 Leithäuser, Joachim G.** *Mappae mundi; die geistige Eroberung der Welt.* Berlin: Safari-Verlag. 1958. 402pp.
Includes reproductions of early maps. Bibliography pp. 397–398.

**84 Miller, Konrad.** *Mappae Arabicae, arabische Welt- und Länderkarten des 9–13 Jahrhunderts in arabischer Urschrift, lateinischer Transkription und Übertragung in neuzeitliche Kartenskizzen. Mit einleitenden Texten hrsg. von Konrad Miller.* Stuttgart: the author, 1926–31. 6 vols.
A comprehensive survey of Arab cartography.

**85 Moir, Arthur Lowndes.** *The world map in Hereford Cathedral by A. L. Moir; the pictures in the Hereford Mappa Mundi by Malcolm Letts; with glossary and bibliography.* 5th revised edn. Hereford: the Cathedral, 1970. 40pp.
Describes and analyses the finest extant medieval *mappa mundi*. Illustrated with details from the map.

**86 Nordenskiöld, Nils Adolf Erik.** *Facsimile-atlas to the early history of cartography, with reproductions of the most important maps printed in the XV and XVI centuries.* Reprinted 1961 by the Klaus Reprint Corporation of New York from the original Swedish edition of 1889. 140pp.
Includes fifty-two large plates and eighty-four illustrations in the text. Investigates the work of Ptolemy as well as early non-Ptolemaic atlases. Includes a

chapter on map projections. A most important reference work for carto-historians, geographers, collectors, historians and laymen.

**87 Rohr, Heinz.** *Die Entwicklung des Kartenbildes Westeuropas zwischen Kanal und Mittelmeer von den ältesten Weltkarten bis Mercator.* Borna-Leipzig: Spezialbetrieb für Dissertationsdruck von R. Noske, 1939. 283pp.
Mainly examines the early cartography of Western Europe up to the time of Mercator but also includes sections on fifteenth- and sixteenth-century world maps.

**88 Sanz López, Carlos.** *Mapas antiguos del mundo (siglos XV–XVI). Reproducidos y comentados por Carlos Sanz.* Madrid: 1962. 157pp.
Reproductions of early world maps with commentary in Spanish.

**89 Sanz López, Carlos.** *Ciento noventa mapas antiguos del mundo de los siglos I al XVIII que forman parte del proceso cartográfico universal.* Madrid: Real Sociedad Geográfica, 1970. 336pp.
Describes and illustrates 190 early world maps that have been an important part of cartographic development. Text in Spanish.

**90 Shirley, Rodney W.** *Mapping of the world: early printed world maps, 1472–1700.* London: Holland Press, 1984. 669pp.
The most comprehensive work on the subject, well illustrated.

**91 Uzielli, Gustavo** and **P. Amat de S. Filippo.** *Mappamondi, carte nautiche, portolani e altri monumenti cartografici specialmente italiani dei secoli XIII–XVII.* Amsterdam: Meridian, 1967. 325pp.
A reprint of a work originally published in Rome in 1882. Includes an annotated list of over 500 world maps and other cartographic material of the thirteenth to seventeenth centuries.

## Regional Mapping

### North America

*Carto-bibliographies and lists published by the Library of Congress*

**92 Phillips, P. Lee.** *A list of maps of America in the Library of Congress. Preceded by a list of works relating to cartography.* Washington, D.C.: Government Printing office, 1901.

**93 Phillips, P. Lee** (compiled under the direction of). *A list of geographical atlases in the Library of Congress, with bibliographical notes.* Washington, D.C.: Government Printing Office. Vols. 1–4. Vols. 5–8 compiled by Clara E. LeGear. 1909–74.
A most important reference work for all librarians and map curators. Covers cartography in general as well as that of America.

**94 LeGear, Clara Egli** (compiler). *United States atlases: A list of national, state, county, city and regional atlases in the Library of Congress.* Washington, D.C.: Government Printing Office, 1950–53. 2 vols.
An indispensable reference for students of American history or of the development of American cartography.

## Other carto-bibliographies and checklists

**95 Cumming, W. P.** *The southeast in early maps, with an annotated check list of printed and manuscript regional and local maps of southeastern North America during the colonial period.* Chapel Hill, North Carolina: University of North Carolina Press, 1962. 284pp.
The map list covers maps of the southeast and local maps south of Virginia and north of the Florida peninsula prior to the American Revolution.

**96 Ganong, W. F.** *Crucial maps in the early cartography and place-nomenclature of the Atlantic coast of Canada.* Toronto: University of Toronto Press, in co-operation with the Royal Society of Canada, 1964. 511pp.

**97 Harlow, Neal.** *The maps of San Francisco Bay, from the Spanish discovery in 1769 to the American occupation.* San Francisco: Book Club of California, 1950. 140pp.
Describes thirty-nine maps in chronological order dating from 1771 to 1847.

**98 Harrisse, Henry.** *Bibliotheca Americana vetustissima; a description of works relating to America, published between the years 1492 and 1551.* Chicago: Argonaut, 1967. 199pp.
One of the most important reference works relating to American discovery.

**99 Harrisse, Henry.** *Découverte et évolution cartographique de Terre-Neuve et des pays circonvoisins, 1497–1501–1769.* Ridgewood, New Jersey: Gregg Press, 1968. 416pp.
A classic reference work on the cartography and discovery of the northeastern coasts.

**100 Jillson, Willard Rouse.** *A checklist of early maps of Kentucky (1673–1825).* Frankfort, Kentucky: Roberts Print Co., 1949. 27pp.
Lists and describes 100 maps of Kentucky.

**101 Karpinski, L. C.** *Bibliography of the printed maps of Michigan, 1804–1880, with a series of over 100 reproductions of maps constituting an historical atlas of the Great Lakes and Michigan.* Lansing, Michigan: Michigan Historical Commission, 1931. 539pp.

**102 Layng, T. E.** *Sixteenth-century maps relating to Canada, a checklist and bibliography.* Ottawa: Public Archives of Canada, 1965. 203pp.

**103 Mason, Sara Elizabeth.** *A list of nineteenth century maps of the state of Alabama.* Birmingham, Alabama: Oxmoor Press, 1973. 256pp.

**104 May, Betty** *et al. County atlases of Canada: a descriptive catalogue.* Ottawa: Public Archives of Canada, 1970. 192pp.

**105 Miller, Ruby M.** *Pennsylvania maps and atlases in the Pennsylvania State University Libraries.* University Park, Pennsylvania: Pennsylvania State University Libraries, 1972. 682pp.

**106 Stephenson, R. W.** *Land ownership maps.* Washington, D.C.: Government Printing Office, 1967. 86pp.

**107 Thompson, Edmund B.** *Maps of Connecticut for the years of Industrial Revolution: 1801–1860.* Windham, Connecticut: Hawthorn House, 1942. 111pp.

**108 U.S. National Archives.** *Civil War maps in the National Archives.* Washington, D.C.: Government Printing Office, 1964. 127pp.
Lists and describes around 8,000 maps and cartographic items relating to the American Civil War.

**109 Verner, Coolie.** *Smith's Virginia and its derivatives: a cartobibliographical study of the diffusion of geographical knowledge.* London: Map Collectors' Circle, 1968. 40pp. (Map Collectors' series No. 45.)

**110 Wheat, Carl I.** *Mapping the Transmississippi west, 1540–1880.* San Francisco: Institute of Historical Cartography, 1957–60. 5 vols.
A detailed carto-bibliography, together with a discussion of the spread of knowledge of the Transmississippi region as revealed by the maps.

**111 Wheat, James Clements** and **Christian F. Brun.** *Maps and charts published in America before 1800, a bibliography.* New Haven, Connecticut: Yale University Press, 1969. 215pp.
A major reference that discusses the entire output of cartographical material printed in America before 1800 and includes an extensive bibliography on pp. 187–207.

**112 Wheat, Carl I.** *The maps of the California gold region, 1848–1857, a bibliocartography of an important decade.* San Francisco: The Grabhorn Press, 1942. 152pp.
Describes over 300 maps in chronological order.

*Selected general works*

**113 Berwick, Jacobo Maria del Pilar Carlos Manuel Stuart Fitz-James,** *10 duque de* (ed.). *Mapas españoles de América, siglos XV–XVII.* Madrid: 1951, 351pp.
A study of the cartographic development of both North and South America during the fifteenth to the seventeenth centuries.

**114 Cumming, William P.** *British maps of Colonial America.* Chicago: University of Chicago Press, 1974. 114pp.
Includes an important inventory of maps held in private collections in Britain.

**115 Fite, Emerson D.** and **Archibald Freeman.** *A book of old maps, delineating American history from the earliest days down to the close of the Revolutionary War.* Cambridge, Massachusetts: Harvard University Press, 1926; New York: Arno Press, 1969. 299pp. Also reprinted, New York: Dover Publications, 1969.
A finely illustrated work in which seventy-four maps provide a survey of the discovery, settlement and growth of America down to the close of the Revolutionary War. The significance of each map is fully explained and there are bibliographical references for each.

**116 Harrisse, Henry.** *The discovery of North America; a critical documentary and historic investigation, with an essay on the early cartography of the New World, including descriptions of 250 maps or globes existing or lost, constructed before the year 1536 . . . and a copious list of the original names of American regions . . . rivers, towns and harbours.* London: H. Stevens & Sons, 1892. 802pp. Reprinted, Amsterdam: N. Israel, 1961.
A thorough investigation into maps concerned with the early history of North America.

**117 Kohl, J. G.** *The Kohl Collection (now in the Library of Congress) of Maps relating to America.* Washington, D.C.: Government Printing Office, 1904. 189pp. Reprint from *Bibliographical Contributions to the Library of Harvard University* No. 19.
Kohl prepared hand-drawn facsimiles of maps relating to America and while these are not faithful reproductions of the original maps they do provide very useful reference material for historians.

**118 Library of Congress.** *Maps and charts of North America and the West Indies, 1750–1789.* Washington, D.C.: Library of Congress, 1981. 495pp.
Describes over 2,000 items, many of them military maps. A valuable reference for historians of North America and also for carto-historians as the maps embody many different types and styles of presentation.

**119 Ristow, Walter W.** *American maps and makers.* Detroit: Wayne State University Press, 1985. 488pp.
A full survey of commercial mapmaking practice and products during the century following the American Revolution.

**120 Schwartz, Seymour I.** and **Ralph E. Ehrenberg.** *The mapping of America.* New York: Abrams, 1980. 363pp.
This important volume covers the history of mapping of the United States from the 1500s to the present time. Includes a bibliography of 120 items.

**121 Thomson, D. W.** *Men and meridians: the history of surveying and mapping in Canada.* Ottawa: R. Duhamel, Queen's Printer, 1966–69. 3 vols.

An important work on cartographic development in Canada that provides useful reference material for historians of cartography and of surveying as well as historians interested in Canada generally.

**122 Wolter, J. A.** 'Source materials for the history of American cartography'. *Bulletin*, Special Libraries Association, Geography and Maps Division, No. 88, pp. 2–16 (1972).
Includes a list of sixty-four references. Mainly aimed at librarians and map curators.

## Central and South America

**123 Adonias, Isa.** *La cartografia da região amazônica: catálogo descritivo; 1500–1961.* Rio de Janeiro: Instituto Nacional de Pesquisas da Amazônia, 1963. 2 vols.
A chronological listing of cartographic material relating to the Amazon region from the period of the discoveries.

**124 Cortes, Vicenta.** *Catálogo de mapas de Colombia.* Madrid: Ediciones Cultura Hispanica, 1967. 356pp.

**125 Guedes, Max Justo.** *Brasil—costa norte; cartografia portuguesa vetustissima. Ed. comemorativa do centenario de Flotilha do Amazonas, 1868–1968.* Rio de Janeiro: Serviço de Documentação Geral da Marinha, 1968. 69pp.
A Portuguese text relating the cartographic history of Brazil's north coast.

**126 Phillips, Philip Lee.** *Guiana and Venezuela cartography.* Washington, D.C.: Government Printing Office, 1898. 53pp.

**127 Servicio de Hidrografía.** *Catálogo del archivo cartografico histórico.* Montevideo: Servicio de Hidrografía, 1956. 393pp.
A list of cartographical material relating to Uruguay.

**128 Uricoechea, Ezequiel.** *Mapoteca colombiana. Colección de los títulos de todos los mapas, planos, vistas etc. relativos a la América Española, Brasil é islas adjacentes.* London: Trübner & Cie, 1860. 215pp.
Maps arranged under separate countries.

## China, Japan and the Pacific Ocean

**129 Hoyanagi, Mutsumi.** 'Re-appreciation of Ino's maps, the first maps of Japan based on actual survey'. Tokyo, *Geographical Reports* of Tokyo Metropolitan University, No. 2, pp. 147–162 (July 1967).
Describes the work of a man who dedicated most of his life to the survey of Japan and who established an important landmark in the history of Japanese cartography.

**130 Jones, Yolande, H. Nelson** and **H. Wallis.** *Chinese and Japanese Maps.* London: The British Library Board, British Museum Publications Ltd., 1974.

An introductory booklet to Chinese and Japanese maps which accompanied an exhibition held in 1974. Includes detailed catalogue of the maps displayed.

**131 Nakamura, Hiroshi.** *East Asia in old maps.* Tokyo: Center for East Asian Cultural Studies; Honolulu: East West Center Press, 1964. 84pp.
Concerned with the period of European exploration in East Asia.

**132 Nanba, M.,** with **N. Muroga** and **K. Unno.** *Old maps in Japan.* Osaka: Sogensha, 1973.
A superbly illustrated history of Japanese maps. Of interest to the general reader as well as carto-historians and students of Japanese history.

**133 Needham, Joseph.** 'Geography and cartography'. In *Science and civilisation in China* (vol. 3, pp. 497–590). Cambridge: Cambridge University Press, 1959. 3 vols.
This is not only an important reference on cartography in China but also includes a valuable section on European cartography. Invaluable for historians, cartographers and geographers.

**134 Quirino, Carlos.** *Philippine cartography, 1320–1899.* 2nd revised edn. Amsterdam: N. Israel, 1963. 140pp.
Includes an important introduction by R. A. Skelton.

**135 Teleki, Pal.** *Atlas zur Geschichte der Kartographie der japanischen Inseln. Nebst dem holländischen Journal der Reise Mathys Quasts und A. J. Tasmans zur Entdeckung der Goldinseln im Osten von Japan i.d.J. 1639. und dessen deutscher Übersetzung.* Originally published in Leipzig in 1909. Authorised reprint, Nedeln, Liechtenstein: Kraus Reprint, 1966. 184pp.
A major reference work on Japanese cartography, which includes an extensive bibliography on pp. 179–184.

**136 Williams, Jack F.** *China in maps, 1890–1960: a selective and annotated cartobibliography.* Asian Studies Center of Michigan State University, 1975. 301pp. (East Asian Series Occasional Paper No. 4.)

**137 Wroth, Lawrence Counselman.** 'The early cartography of the Pacific'. Bibliographical Society of America, *Papers,* **38**(2): 87–268 (1944).
Describes over 100 maps ranging from Ptolemy's time to 1798.

**138 Zerlik, Alfred.** *P. Xaver Ernbert Fridelli, Chinamissionar und Kartograph aus Linz.* Linz: Oberösterreichischer Landesverlag, 1962. 68pp.
Discusses the work of the eighteenth-century Austrian missionary–cartographer, Fridelli, and includes listings of early-eighteenth-century maps of China.

## Ceylon (Sri Lanka)

**139 Sinnatamby, J. R.** *Ceylon in Ptolemy's geography.* Colombo: 1968. 73pp.
Sinnatamby looks at some of the more general aspects of *Geographia* as well as dealing with the identification of Ptolemy's placenames in Ceylon.

# India

**140 Gole, Susan.** *Early maps of India.* New Delhi: Sanskriti, in association with Arnold/Heinemann Publishers (India), 1976. 126pp.
A well-illustrated introduction to the history of mapmaking as it concerns India.

# Australasia

**141 Sanz López, Carlos.** *Cartografía histórica de los descubrimientos australes.* Madrid: Impr. Aguirre, 1967. 96pp. (Publicaciónes de la Real Sociedad Geográfica, Serie B, No. 471.)
Discusses the evolution of the legendary 'Terra Australis Incognita' as shown on early maps.

**142 Tooley, Ronald Vere.** *Early maps of Australia, the Dutch period; being examples from the collection of R. V. Tooley with bibliographical notes.* London: The Map Collectors' Circle, 1965. 27pp. (Map Collectors' Series No. 23.)
Tooley deals with the early Dutch discoveries before Tasman; Tasman's voyages, 1642–44; and the cartographic representation of Australia before 1770.

# Africa

**143** *Africa: maps and plans in the Public Record Office.* London: HMSO, 1982.
The third volume in the catalogue of maps and plans in the Public Record Office contains a wealth of information on the exploration and survey of the continent and is a valuable reference tool for African scholars, librarians and carto-historians.

**144 Klemp, Egon** (foreword by). *Africa on maps dating from the twelfth to the eighteenth century.* Leipzig: Edition Leipzig, 1968.
This magnificent volume consists of seventy-seven facsimiles of early maps bound together with an accompanying booklet. German text. Contains material of wider interest than Africa alone but is essential study material for all Africanists and carto-historians.

**145 Lane-Poole, E. H.** *The discovery of Africa: a history of the exploration of Africa as reflected in the maps in the collection of the Rhodes–Livingstone Museum.* Livingstone: Rhodes–Livingstone Museum, 1950. 28pp. (Rhodes–Livingstone Occasional Papers New Series No. 7).
Lists fifty-six maps dating from 1478 to 1857.

**146 Mota, Avelina Teixeira da.** *A cartografia antiga da Africa Central e a travessia entre Angola e Moçambique, 1500–1860.* Lourenço Marques: Sociedade de Estudos de Moçambique, 1964. 225pp.
Includes extensive bibliography on pp. 247–255.

**147 Tooley, Ronald Vere.** *Collector's guide to maps of the African continent and southern Africa.* London: Carta Press, 1969. 132pp.

An important reference book that brings together work already published in several volumes of the Map Collectors' Series.

**148 Traversi, Carlo.** *Storio della cartografia coloniale italiana.* Rome: Istituto Poligrafico dello Stato, 1964. 294pp.
The history of Italian mapmaking in former Italian colonies on the African continent. Extensive bibliography on pp. 247–257.

**149 Youssouf Kamal,** *Prince. Monumenta cartographica Africae et Aegypti.* Cairo: privately published, 1926–51. 5 vols.
Includes hundreds of facsimiles of early maps. Unfortunately published only in a limited edition of seventy copies which were presented to selected institutions throughout the world.

## The Middle East

**150 Alavi, S. M. Ziauddin.** *Arab geography in the ninth and tenth centuries.* Aligarh: Department of Geography, Aligarh Muslim University, 1965. 134pp.
Includes discussions on Arabic cartography and a good bibliography. For cartohistorians and geographers.

**151 Haifa Maritime Museum.** *Old maps of the land of Israel.* Haifa: Haifa Maritime Museum, 1963. 57pp.
Describes eighty-five maps of which some are reproduced in colour. Hebrew and English text.

**152 Miller, K.** *Mappae Arabicae, arabische Welt- und Länderkarten des 9–13 Jarhunderts in arabischer Urschrift, lateinischer Transkription und Übertragung in neuzeitliche Kartenskizzen. Mit einleitenden Texten hrsg. von Konrad Miller.* Stuttgart. 1926–31. 6 vols contained in 3 portfolios.
An important work by a German scholar on Arabic mapping. The contributions of Idrisi, and particularly his world map of 1154, are stressed.

**153 Vilnay, Zev.** *The Holy Land in old prints and maps.* 2nd edn. Jerusalem: R. Mass, 1965. 296pp.
A lavishly illustrated work with many reproductions of early maps. Contains bibliographies.

## Russia

**154 Bagrow, Leo.** *A history of the cartography of Russia up to 1600.* Ed. by Henry W. Castner. Wolfe Island, Ontario: The Walker Press, 1975. 140pp.

**155 Bagrow, Leo.** *A history of Russian cartography up to 1800.* Ed. by Henry W. Castner. Wolfe Island, Ontario: The Walker Press, 1975. 312pp.
This and the previous volume by Bagrow contain extensive bibliographies. The two works are edited from translated manuscripts written by the late Leo Bagrow

and constitute an important reference source for librarians and carto-historians as well as students of Russian geography.

**156** Cartographica Monograph No. 13. *Essays on the history of Russian cartography 16th to 19th centuries.* Toronto: York University, 1975. 145pp.

**157 Fel', Sergei Efimovich.** *Kartografiia Rossii XVIII veka* [Eighteenth-century Russian cartography]. Moscow: Izdatelstvo Geodezicheskoi Lit-Ry, 1960. 226pp.
Russian text summarizing the eighteenth-century history of Russian mapmaking.

**158 Goldenberg, Leonid A.** *Russian maps and atlases as historical sources.* Toronto: York University, 1971. 76pp. (Cartographica Monograph No. 3.)

**159 Ostrowski, Wiktor.** *The ancient names and early cartography of Byelorussia: material for historical research and study.* Revised edn. London: Ostrowski, 1970. 20pp.
Includes material concerning Russian cartographical history in general.

**160 Shibanov, Fyodor A.** *Studies in the history of Russian cartography Part I.* Toronto: York University, 1975. 101pp. (Cartographica Monograph No. 14.)

**161 Shibanov, Fyodor A.** *Studies in the history of Russian cartography Part II.* Toronto: York University, 1975. 86pp. (Cartographica Monograph No. 15.)

## European countries

### Belgium
**162 Wauwermans, Henri Emmanuel.** *Histoire de l'école cartographique belge et anversoise du XVIe siècle.* Brussels: Institut National de Géographie, 1895. 2 vols.
Emphasizes the importance of the sixteenth-century school of mapmaking established in Belgium, and particularly in Antwerp.

### Czechoslovakia
**163 Kuchař, Karel.** *Early maps of Bohemia, Moravia and Silesia.* Prague: 1961. 74pp.
A translation by Zdeněk Safařik which describes the life and work of important Central European mapmakers.

### Denmark
**164 Bramsen, Bo.** *Gamle Danmarkskort; en historisk oversigt med bibliografiske noter for perioden 1570–1770* [Old Danish maps; a history with bibliographical notes for the period 1570–1770]. Copenhagen: Politikens Forlag, 1952. 159pp.
The most comprehensive cartographic history of Denmark. Very well-illustrated with many reproductions of early maps.

**165 Kejlbo, Ib.** *Historisk kartografi.* Copenhagen: Dansk Historisk Faelles forenings Håndbøger, 1966. 83pp.
A brief history of cartography in general, but with the emphasis on Scandinavia. Includes bibliography.

**166 Nørlund, Niels Erik.** *Danmarks kortlaegning, en historisk fremstilling; udg. med støtte af Carlsbergfondet* [Danish mapmaking, a historical representation, published with the aid of the Carlsberg Foundation]. Copenhagen: Munksgaard, 1943. 77pp. (Geodaetisk Instituts Publikationer 4.)
Specifically concerned with Denmark but also looks at the mapmaking of Scandinavia as a whole. Very well illustrated.

### France

**167 Armée, Service Géographique.** *La carte de France, 1750–1898. Etude historique par le Colonel Berthaut, chef de la Section de Cartographie.* Paris: Imprimerie du Service Géographique, 1898–99. 2 vols.
A comprehensive treatment of the official mapmaking of France from 1750.

**168 Pedley, Mary Sponberg.** 'The map trade in Paris, 1650–1825'. *Imago Mundi*, **33**: 33–45 (1981).
An important article that looks at the location of those engaged in cartographical pursuits from the time of Sanson to the first quarter of the nineteenth century, when lithography superseded copperplate engraving. Bibliography pp. 43–45.

### Germany

**169 Bonacker, Wilhelm.** *Grundriss der fränkischen Kartographie des 16. und 17. Jahrhunderts.* Würzburg: Freunde Mainfränkischer Kunst und Geschichte; Auslieferung, Buchdr. K. Hart. Volkach vor Würzburg, 1959. 75pp.
A discussion of the cartography of Franconia during the sixteenth and seventeenth centuries.

**170 Durand, Dana B.** *The Vienna–Klosterneuburg map corpus of the fifteenth century: a study in the transition from medieval to modern science.* Leiden: E. J. Brill, 1952. 510pp.
An examination of the fifteenth-century beginnings of German cartography. Extensive bibliography on pp. 290–321.

**171 Geerz, F.** *Geschichte der geographischen Vermessungen und der Landkarten Nordalbingiens vom Ende des 15. Jahrhunderts bis zum Jahre 1859.* Berlin: Commissionsdebit, 1859. 281pp.
Looks at the mapping of the Lower Rhine region including Schleswig–Holstein and Denmark.

**172 Oehme, Ruthardt.** *Die Geschichte der Kartographie des deutschen Süd-westens. Hrsg. von der Kommission für Geschichtliche Landeskunde in Baden-Württemberg.* Konstanz: J. Thorbecke, *c.* 1961. 168pp.
An examination of the development of mapmaking of southwest Germany from the thirteenth to nineteenth centuries. Contains a bibliography.

## Greece

**173 Clutton, Elizabeth.** 'Some seventeenth century images of Crete: a comparative analysis of the manuscript maps of Grancesco Basilicata and the printed maps by Marco Boschini'. *Imago Mundi*, **34**: 48–65 (1982).
Has a useful list of references on pp. 64–65.

## Iceland

**174 Sigurdsson, H.** *Kortasaga Islands fra lokum 16 aldar til 1848* [Cartographic history of Iceland from the sixteenth century to 1848]. Reykjavik. Vol. 1, 1971; vol. 2, 1978.
Written in Icelandic with a summary in English.

## Italy

**175 Aliprandi, L., G. Aliprandi** and **M. Pomella.** *Le Grandi Alpi nella cartografia dei secoli passati 1482–1865.* Ivrea: Priuli & Verlucca, July 1974. 469pp.
A superbly illustrated examination of the history of mapping the Great Alps and the passes between the Aosta Valley, Savoy and the Valois and the Gran Paradiso. Text in Italian, English and French. A most important study and reference for carto-historians and Italian scholars.

**176 Almagià, Roberto.** *Monumenta Italiae cartographica. Riproduzioni di carte generali e regionali d'Italia dal secolo XIV al XVII. Raccolta e illustrate da Roberto Almagià.* Florence: Istituto Geografico Militare, 1929. 88pp.
Almagià evaluates and describes the work of Italian mapmakers and provides bibliographical details.

**177 Destombes, Marcel.** *Les cartes de Lafréri et assimilées (1532–1586) du Département des Estampes de la Bibliothèque Nationale.* Paris: Nouvelles de l'Estampe, 1970.
Lists over 200 maps and atlases by Lafréri together with similar maps and plans. Bibliography.

## Netherlands

**178 Fockema Andreae, S.J.** and **B. van 't Hoff.** *Geschiedenis der Kartografie van Nederland, van den Romeinschen tijd tot het midden der 19de eeuw, door S. J. Fockema Andreae, met medewerking van B. van 't Hoff.* The Hague: Martinus Nijhoff, 1947. 127pp.
Looks at the development of the cartographical presentation of the Netherlands and sets it against its historical and cultural background. Has a summary in English.

**179 Koeman, C.** *Collections of maps and atlases in the Netherlands: their history and present state.* Leiden: E. J. Brill, 1961. 301pp. (*Imago Mundi* Supplement 3.)
An important reference work in Dutch and English that provides source material for the study of the history of cartography in the Netherlands. Includes a list of map collections in the Netherlands at the present time. Formulates a schematic

outline of the bibliography of atlases published in the Low Countries before 1800. Studies the character of map collecting through the centuries.

**180 Koeman, C.** (ed.). *Atlantes Neerlandici, bibliography of terrestrial, maritime and celestial atlases published in the Netherlands up to 1880.* Amsterdam: Theatrum Orbis Terrarum, 1969. 5 vols.
An indispensable work for librarians and curators of collections dealing with the history of printing and publishing, geography and cartography, history and art. Describes 1,017 editions of atlases by 240 publishers and 680 volumes of maritime atlases. Provides biographical data on mapmakers and publishers.

**181 Koeman, C.** *Hanleiding voor de studie van de topografische kaarten van Nederland 1750–1850.* Groningen: J. B. Wolters, 1963. 120pp.
A guide to mapping in the Netherlands between 1750 and 1850. Useful reference work for geographers, historians, surveyors and planners.

## Norway

**182 Engelstad, Sigurd.** *Norge i kart gjennom 400 år. Med opplysninger om dem som utformet kartbildet* [Norway in maps over a 400-year period. With illustrations of outstanding maps]. Oslo: J. W. Cappelens Antikvariat, 1952. 112pp.
Illustrates and describes the development of Norwegian maps from earliest times to the present. Includes an annotated bibliography of some 200 items and a list of 471 maps. In Norwegian but with an English foreword.

## Poland

**183 Buczek, Karol.** *The history of Polish cartography from the fifteenth to the eighteenth century.* Warsaw: Ossolineum, 1966. 136pp. 3rd edition: Amsterdam: Meridian, 1982.
An introduction to mapmaking in Poland. There are sixty maps in a pocket at the end of the volume.

## Portugal

**184 Cortesão, Armando.** *Cartografìa e cartógrafos portugueses dos séculos XV e XVI.* Lisbon: Edição da 'Seara Nova', 1935. 2 vols.
An outstanding study of Portuguese portolan charts and their makers. An important reference work with many bibliographical references and numerous facsimile reproductions.

**185 Cortesão, Armando.** *History of Portuguese cartography.* Lisbon: Junta de Investigações do Ultramar, 1969–71. Vol. 1, 1969; vol. 2, 1971.
Volume 1 looks at mapmaking from the earliest times to the fourteenth century while vol. 2 discusses the fourteenth and fifteenth centuries with two chapters devoted to the history of astronomical navigation.

**186 Cortesão, Armando** and **Avelino Teixeira da Mota.** *Portugaliae monumenta cartographica.* Lisbon: Portuguese government, 1960–62. 6 vols.
One of the most important and most finely produced works on the history of

cartography. Includes numerous reproductions of early maps. Lists Portuguese mapmakers prior to the eighteenth century. Text in Portuguese and English.

**187 Denucé, Jean.** *Les origines de la cartographie portugaise et les cartes des Reinel.* Ghent: E. van Goethem, 1908. 136pp.
As well as studying the early mapping by Portuguese mapmakers in general, Denucé pays particular attention to the work of the two Reinels: Pedro and Jorge.

### Sweden

**188 Bratt, Einar.** *En kronika om kartor över Sverige.* Stockholm: Generalstabens Lithografiska Anstalt, 1958. 131pp.
A general treatment of the history of maps of Sweden from earliest times to the present. Swedish text.

**189 Dahlgren, Per Johan,** with **Herman Richter.** *Sveriges sjökarta* [Swedish sea-charts]. Lund: H. Ohlssons boktryckeri, 1944. 413pp. (Statens Sjöhistoriska Museum Handlingar 1.)
Deals specifically with hydrographic surveys of Sweden. Has an English summary.

### Switzerland

**190 Blumer, Walter.** *Die topographischen Karten des Kantons Glarus.* Einsiedeln: In Kommission bei Benziger, 1950. 44pp.
Looks at the mapping of Glarus from 1496. Contains a bibliography.

**191 Grob, Richard.** *Geschichte der schweizerischen Kartographie.* Berne: Kummerley & Frey, 1941. 194pp.
Particularly useful on the sixteenth to eighteenth centuries but also looks briefly at earlier periods.

**192 Imhof, Eduard.** *Die ältesten Schweizerkarten. Mit einem Faksimile der ältesten gedruckten Schweizerkarte von 1513.* Zürich: Orrell Füssli Verlag, 1939. 15pp.
A brief look at some early Swiss maps.

**193 Weisz, Leo.** *Die Schweiz auf alten Karten. Mit Geleitw. und einem kartographisch-technischen Anhang von Ed. Imhof.* Zürich: Buchverlag der Neuen Zürcher Zeitung, 1969. 247pp.
A well-illustrated history of Swiss cartography and of Swiss maps. Bibliography pp. 242–245.

### United Kingdom: Carto-bibliographies

**194 Chubb, Thomas.** *The printed maps in the atlases of Great Britain and Ireland, 1579–1880.* Originally published in 1927 by the Homeland Association. Reprint, Folkestone: Wm. Dawson & Sons, 1974.
For many years 'Chubb' was the standard work on that distinctive cartographic

form, the English county atlas. Still one of the foremost reference works for historians, map curators, librarians, map collectors and all students and users of early maps.

**195 Skelton, R. A.** *County atlases of the British Isles, 1579–1850. A bibliography compiled by R. A. Skelton. 1579–1703.* London: Carta Press, 1970. 262pp.
The most important carto-bibliography of county atlases for the period 1579–1703. Incorporates recent studies and discoveries. The definitive reference work for its period. Since Dr. Skelton's death his work has been continued by Hodson (*see* p. 29, note added in proof).

**196 Rodger, Elizabeth M.** (compiler). *The large-scale county maps of the British Isles, 1596–1850, a union list.* 2nd revised edn. Oxford: Oxford University Press, 1972. 56pp.
Lists county maps issued separately up to 1850 at a scale between half an inch and three inches to one mile. With 843 entries. The only detailed reference source for large-scale county maps of the British Isles.

**197 Smith, David.** *Antique maps of the British Isles.* London: Batsford, 1982. 243pp.
Designed mainly for collectors of early maps and for professional dealers. The early chapters provide a brief introduction to British mapping while the main body of the book gives detailed descriptions of the maps as a means to identification.

## United Kingdom: selected general works

**198 Harley, J. B.** *Maps for the local historian: a guide to the British sources.* London: National Council of Social Service, for the Standing Conference for Local History, 1972. 80pp.
A short but extremely useful reference work on many types of British map which is of prime assistance to local historians, collectors, dealers, carto-historians and map curators.

**199 Lynam, Edward.** *British maps and map-makers.* London: Collins, 1947. 48pp.
A useful introduction to British maps by the then superintendent of the Map Room in the British Museum. Of interest to the general reader as well as to anyone with a professional interest in maps.

**200 Royal Scottish Geographical Society.** *The early maps of Scotland to 1850.* Edinburgh: The Royal Scottish Geographical Society, 1973. 243pp.
The first edition of this volume was published in 1934 to mark the jubilee of the Society. Much additional information has now been added, together with 'A history of Scottish maps' by D. G. Moir. Includes much biographical information on cartographers and publishers.

**201 North, F. G.** *The map of Wales (before 1600 AD).* Cardiff: National Museum of Wales and the Press Board of the University of Wales, 1935. 69pp.

A narrative, analytical account of the early mapping of Wales. Useful for historians, geographers and carto-historians.

## Urban Plans; History and Development

The mapping of towns and cities is a field that has been relatively neglected by authors and there is at present no major work available which is solely devoted to the topic. The following is a selective list of some important reference works and introductory texts.

**202 Adonias, Isa,** with **Marta M. Gonsalves** and **Yolette Soares de Miranda.** *Catálogo de plantas e mapas da cidade do Rio de Janeiro.* Rio de Janeiro: Seção de Publicações, 1966. 171pp.
Lists over 200 plans of the city.

**203 Danckaert, L.** *Plans et vues de dix-neuf villes belges. Catalogue de l'exposition.* Brussels: Bibliothèque Royale, 1968. 95pp.
The catalogue describes 107 plans which cover nineteen towns in Belgium. Of major interest to urban planners and historians.

**204 Darlington, Ida** and **James Howgego.** *Printed maps of London circa 1553–1850.* London: Philip, 1964. Reprint, Folkestone: Wm. Dawson & Sons, 1978. 257pp.
A comprehensive guide for urban and map historians to the printed maps and plans of the city that have survived from the mid sixteenth century. A standard reference work. Includes information on canals and railways as delineated on the plans of the city.

**205 Fordham, Angela.** *Town plans of the British Isles.* London: Map Collectors' Circle, 1965. (Map Collectors' series No. 22.)

**206 Glanville, Philippa.** *London in maps.* London: The Connoisseur, 1972. 212pp.
A superbly illustrated work with sixty-nine plates. Suitable for the general reader as well as being a useful reference for historians, librarians, planners and urban geographers.

**207 Haskell, Daniel C.** (ed.). *Manhattan maps: a co-operative list.* New York: New York Public Library, 1931. 128pp.
Describes nearly 2,000 maps and plans, ranging over the period 1600–1930. Chronologically arranged.

**208 Hebert, John R.** *Panoramic maps of Anglo-American cities: a checklist of maps in the collections of the Library of Congress, Geography and Map Division.* Washington, D.C.: Superintendent of Documents, Government Printing Office, 1974.
A checklist, available free from the above address, that provides relevant data on 1,117 panoramic maps of cities in the United States and Canada. Includes an

article on the panoramic map business in America and the individuals and firms involved. An important reference work for urban planners, cartographers and urban historians.

**209 Hodson, D.** *Maps of Portsmouth before 1801.* Portsmouth: City of Portsmouth, 1978. 169pp.
A definitive carto-bibliography of Portsmouth maps and plans that would serve as the ideal model for anyone attempting to prepare a similar work for other British cities. Describes 370 maps, all but five of which are in manuscript. Important for historians, planners, sociologists.

**210 Hyde, R.** *Ward maps of the City of London.* London: Map Collectors' Circle, 1967. (Map Collectors' series No. 38.)

**211 Hyde, R.** *Printed maps of Victorian London 1851–1900.* Folkestone: Wm. Dawson & Sons, 1975.
A most important, authoritative reference work that completes the story of London maps begun by Darlington and Howgego [204]. An indispensable work for librarians, map curators and urban historians.

**212 Lübbecke, Fried.** *Das Antlitz der Staat; nach Frankfurts Plänen von Faber, Merian und Delkeskamp 1552–1864.* Frankfurt am Main: W. Kramer, 1952. 156pp.
Includes biographical sketches of Faber, Merian and Delkeskamp as well as other engravers of Frankfurt plans.

**213 Map Collectors' Circle.** *North American city plans.* London: Map Collectors' Circle, 1965. (Map Collectors' series No. 20.)
Includes numerous illustrations of early plans.

**214 Merian, Matthaeus.** *Flores Electi ex Matthaei Meriani Topographiae Germaniae XXIV volumibus. Ausgewählte Blätter aus den vierundzwanzig Bänden deutscher Topographie des Matthäus Merian (1593–1650).* Frankfurt am Main: Deutsche Lufthansa AG, in co-operation with Graphischen Grossbetrieb K. G. Lohse.
A magnificently illustrated work with facsimile reproductions of Merian's plans of German towns. Merian's original text in German is given with translations into English, French and Spanish. A valuable insight into the work of one of the most significant of early makers of urban plans and views. Recommended to urban historians, planners and sociologists as well as to the general reader.

**215 Merian, Matthaeus.** *Die schönsten Städte Alt-Österreichs. Aus der Archontologia-cosmica und den Topographien, mit einer Einleitung von Bruno Grimschitz.* Hamburg: Hoffmann & Campe, 1963. xv pp.
A superb selection of Merian's work, featuring forty-one plates. Of interest to all whose researches involve the study of Austrian cities. Has an introduction in German on the life of Merian.

**216 Otness, Harold M.** *Index to nineteenth century city plans appearing in guidebooks.*

Santa Cruz, California: Western Association of Map Libraries, 1980. 84pp. (Occcasional Paper No. 7.)
This index covers 1,800 plans of around 600 places in 164 guidebooks. Places are alphabetically arranged and there is an introduction to nineteenth-century guidebooks, publishers and travellers. Short bibliography. Good reference aid for map curators.

**217 Otness, Harold M.** *Index to early twentieth century city plans appearing in guidebooks.* Santa Cruz, California: Western Association of Map Libraries, 1978. 91pp. (Occasional Paper No. 4.)
Covers more than 2,000 plans of places. The arrangement is alphabetical. Has an introduction that discusses the history and development of guidebook publishing since Baedeker issued its first guide in 1839. These two volumes by Otness provide a valuable guide to material that is normally scattered under a variety of subject headings and which is an important reference source for students of urban planning, urban geographers and historians, cartographers and others.

**218 Phillips, Philip Lee.** *A descriptive list of maps and views of Philadelphia in the Library of Congress, 1683–1865.* Philadelphia: Geographical Society of Philadelphia, 1926. 91pp.
Lists 490 maps from a collection that is particularly noted for its material on the Revolutionary War. A reference aid for urban and military historians.

**219 Reps, J. W.** 'Boston by Bostonians: the printed plans and views of the colonial city by its artists, cartographers, engravers and publishers'. In *Boston prints and printmakers, 1670–1775*, pp. 3–56. Charlottesville: Colonial Society of Massachusetts. University Press of Virginia, 1973.

**220 Robinson, W. W.** *Maps of Los Angeles from Ord's survey of 1849 to the end of the boom of the eighties.* Los Angeles: Dawson's Bookshop, 1966. 87pp.

**221 Romanelli, G.** and **S. Biadene.** *Venezia Pianti e Veduti i Catalogo del fondo cartografico a stampa.* Venice: Museo Correr, April 1982. 118pp.
Well-illustrated catalogue of an extensive exhibition of plans of Venice held in the Museo Correr in 1982. Valuable reference work on the city for librarians, map curators, urban planners etc.

**222 Samkalden, I.** (introduction by). *Opkomst en bloei van het Noordnederlandse stadsgezicht in de 17de eeuw* [The Dutch cityscape in the seventeenth century and its sources]. Amsterdam: Amsterdams Historisch Museum; Toronto: Art Gallery of Toronto, 1977. 272pp.
An exhibition catalogue of paintings, drawings, prints, maps and plans relating to seventeenth-century Holland. Includes an interesting and well illustrated section on 'Kaarten en profielen' (maps and profiles) on pp. 100–119. An important aid to forming a picture of the appearance and character of Dutch cities in the seventeenth century. For planners, historians, cartographers and city dwellers of today. The exhibition was shown in both Amsterdam and Toronto and this collaboration between the museums of the respective cities is reflected in

the Dutch and English texts of the introductory essays (including a particularly interesting one on 'Maps, books and prints: the topographical tradition in the northern Netherlands' by Boudewijn Bakker) and of the notes to individual exhibits.

# Hydrography and Nautical Charts

### Bibliography; catalogues

**223 Foncin, Myriem, Marcel Destombes** and **Monique de la Roncière.**
*Catalogue des cartes nautiques sur vélin conservées au Département des Cartes et Plans.*
Paris: Bibliothèque Nationale, 1963. 317pp.
Lists and describes over 180 manuscript charts on vellum. Includes a comprehensive bibliography of literature concerned with portolan charts. An important reference for historians of sea-charts and mapmaking, for librarians and for hydrographers.

**224 Mariners' Museum**, Newport News, Virginia. *Catalog of maps, ships' papers and logbooks.* Boston: G. K. Hall, 1964. 505pp.
Reproduced from cards in the catalogue of the Museum, which houses a collection of over 1,600 maps.

**225 National Maritime Museum**, Greenwich. *Catalogue of the Library:* vol. iii, *Atlases and cartography.* London: HMSO, 1971. 2 vols.
A most important reference catalogue for students of hydrography, carto-historians, etc.

### Major works

**226 Anthiaume, A.** *Cartes marines, constructions navales, voyages de découverte chez les Normands, 1500–1650, par l'abbé A. Anthiaume.* Paris: E. Dumont, 1916. 2 vols.
Concerned with French portolan charts and French discoveries in the New World.

**227 Barbosa, Antonio.** *Novos subsídios para a história da ciência náutica portuguesa da época dos descobrimentos.* Porto: Instituto para a Alta Cultura, 1948. 332pp.
An important work, with Portuguese text, about the navigational charts of the fourteenth to sixteenth centuries, the age of the great discoveries. Contains a bibliography of some 100 titles.

**228 Blewitt, Mary.** *Surveys of the seas: a brief history of British hydrography.*
Foreword by Archibald Day; appendix on ships and instruments by G. P. B. Naish. London: MacGibbon & Kee, 1957. 168pp.
Illustrated with reproductions of early charts.

**229 Cortesão, Armando.** *Cartografia e cartógrafos portugueses dos séculos XV e XVI.*
Lisbon: Edição da 'Seara Nova', 1935. 2 vols.
A comprehensive work on Portuguese portolan charts and their makers. Of importance to historians, carto-historians, hydrographers, etc.

**230 Dawson, Llewellyn S.** *Memoirs of hydrography, including brief biographies of the principal officers who have served in H.M. Naval Surveying Service between the years 1750 and 1885.* Eastbourne: H. W. Keay, 1883–85. Facsimile reprint, London: Cornmarket Press, 1969. 2 vols. in one.
Contains biographical details of hydrographers from many countries and constitutes a useful reference for librarians and students of hydrographic history, surveying, etc.

**231 Day, Archibald.** *The Admiralty Hydrographic Service, 1795–1919.* London: HMSO, 1967. 378pp.
A definitive history written by a former Hydrographer of the Navy.

**232 Howse, Derek** and **Michael Sanderson.** *The sea chart: an historical survey based on the collections in the National Maritime Museum.* Newton Abbot, Devon: David & Charles, 1973. 144pp.
A superbly illustrated account with excellent bibliographies. Of general interest as well as for the specialist historian of cartography, sea charts or surveying.

**233 Kretschmer, K.** *Die italienischen Portolane des Mittelalters; ein Beitrag zur Geschichte der Kartographie und Nautik.* Hildesheim: G. Olms, 1962. 688pp. A reprint of a 1909 publication by Mittler, Berlin.
An important study of portolan charts, particularly those made by the Italian school centred on Genoa.

**234 La Guardia Trias, R. A.** *La aportación científica de Mallorquines y Portuguéses a la cartografía naútica en los siglos XIV al XVI.* Madrid: Consejo Superior de Investigaciónes Científicas, Instituto Histórico de Marina, 1964. 72pp.
A specialized study of portolan charts made in Majorca and later in Portugal.

**235 Nordenskiöld, Nils Adolf Erik.** *Periplus: an essay on the early history of charts and sailing-directions; translated from the Swedish original by Francis A. Bather.* Stockholm: Norstedt, 1897. Reprint, New York: Burt Franklin Research, 1967.
An important reference work by the explorer and carto-historian, Nordenskiöld, which covers not only portolan charts but also *mappae mundi* and regional maps. The maps are described in chronological order. The index is comprehensive and there are sixty facsimile reproductions. An indispensable work for historians of cartography and hydrography.

**236 Rey Pastor, Julio** and **Ernesto Garcia Camarero.** *La cartografía mallorquina.* Madrid: Departamento de Historia y Filosofía de la Ciencia, 'Instituto Luis Vivas', Consejo Superior de Investigaciónes Científicas, 1960. 207pp.
Discusses the important school of cartography in Majorca headed by the Cresques family, which produced fine charts from the fourteenth to the sixteenth centuries. Biographical sketches. Bibliography pp. 171–191.

**237 Ritchie, G. S.** *The Admiralty chart: British naval hydrography in the nineteenth century.* London, Sydney and Toronto: Hollis & Carter, 1967; New York: Elsevier Publishing Co., 1967. 388pp.

Rear Admiral Ritchie draws on his experiences as a marine naval surveyor and hydrographer to describe what lies behind the compilation of the Admiralty charts. There is a useful bibliography on pp. 373–379.

**238  Robinson, Adrian H. W.** *Marine cartography in Britain: a history of the sea chart to 1855, with a foreword by John Edgell.* Leicester: Leicester University Press, 1962. 222pp.
A valuable, well-illustrated scholarly study of British nautical cartography and the ways in which its development is linked with the science of marine surveying. Covers the period from the reign of Henry VIII to the mid nineteenth century. With biographical notes on some sixteenth-century surveyors and chartmakers. Lists the products of individual chartmakers.

**239  Uzielli, Gustavo** and **P. Amat de S. Filippo.** *Mappamondi, carte nautiche, portolani e altri monumenti cartografici specialmente italiani dei secoli XIII–XVII.* Amsterdam: Meridian, 1967. 325pp.
Lists over 500 world maps, nautical charts, portolan charts and other Italian cartographic works from the thirteenth to the seventeenth centuries. There is a good bibliography on pp. 303–312. With an introductory essay.

**240  Waters, D. W.** *The rutters of the sea: the sailing directions of Pierre Garcie.* New Haven: Yale University Press, 1967. 478pp.
A study of the early sixteenth-century 'rutters', the forerunners of the nautical atlases. Bibliography pp. 469–473.

**241  Youssouf Kamal.** *Hallucinations scientifiques (les portulans).* Leiden: E. J. Brill, 1937. 95pp.
The 'hallucinations' are connected with the idea that the early portolans bear witness to Arabic influence and that there is any definite knowledge about their origins.

# Survey and Surveying Instruments: Development

### Selected works

**242  Bachmann, E.** *Wer hat Himmel und Erde gemessen? Von Erdmessungen, Land-karten, Polschwankungen, Schollenbewegungen, Forschungsreisen und Satelliten.* Munich: Auflag Thun, 1968. 296pp.
This German text discusses the history of surveying and mapping from early origins to modern times. Has a bibliography on p. 296. For students of survey and cartography, and carto-historians.

**243  Bedini, Silvio A.** *Early American scientific instruments and their makers.* Washington, D.C.: Museum of History and Technology; Smithsonian Institution; U.S. Government Printing Office, 1964. 184pp. (U.S. National Museum Bulletin 231.)
Includes references to, and illustrations of, early surveying instruments and discusses surveyors, cartographers and globemakers.

**244 Cazier, Lola.** *Surveys and surveyors of the public domain, 1785–1975.* Washington, D.C.: Government Printing Office, 1976. 228pp.
Largely concerned with the surveyors who were engaged on the rectangular surveys that started on a regular basis with the Land Ordinance of 1785. The books is in narrative style with anecdotes, its aim being 'to be used as an aid in training cadastral surveyors in the application of surveying principles'.

**245 Dilke, O. A. W.** *The Roman land surveyors: an introduction to the agrimensores.* Newton Abbot: David & Charles, 1971. 260pp.
An authoritative work on the surveying methods of the Romans by the foremost authority on the subject.

**246 Dilke, O. A. W.** 'Maps in the treatises of Roman land surveyors'. *Geographical Journal,* **cxxvii**: 417–426 (1961).

**247 Dilke, O. A. W.** 'Illustrations from Roman surveyors' manuals'. *Geographical Journal,* **xxi**, 9–29 (1967).
In this and the foregoing paper Professor Dilke discusses the pictorial maps that so often illustrated the surveying manuals of the Romans. Of interest to carto-historians, surveyors, geographers and students of the Roman Empire.

**248 Johnston, Frederick W.** *Knights and theodolites: a saga of surveyors.* Sydney: Edwards & Shaw, 1962. 232pp.
Discusses generations of Western Australian land surveyors in a non-technical way, making the book suitable for laymen as well as for professional surveyors and cartographers.

**249 Pattison, William D.** *Beginnings of the American rectangular survey system, 1784–1800.* Chicago: University of Chicago, 1957. 248pp. (Department of Geography Research Paper 50.)
A report on the years of the land survey system in the United States. Good bibliography.

**250 Richeson, Allie Wilson.** *English land measuring to 1800: instruments and practices.* Cambridge, Massachusetts: The Society for the History of Technology and the MIT Press, 1966. 214pp. (Society for the History of Technology Series of Monographs, No. 2.)
An excellent account of the development of land surveying in England from pre-Roman times to 1800. Emphasizes the development of surveying as a science and the construction of instruments. Has a good bibliography on pp. 189–207. A most important and unique work for historians, surveyors, cartographers, geographers and earth scientists.

**251 Royal Institution of Chartered Surveyors.** *Five centuries of maps & mapmaking. An exhibition at 12 Great George Street, Westminster.* London: RICS, 1953. 127pp.
Uses 999 maps dating from 1504 to 1945 to illustrate the development of surveying and mapmaking as a science in Britain.

**252 Thompson, F. M. L.** *Chartered surveyors: the growth of a profession.* London: Routledge & Kegan Paul, 1968. 400pp.
A detailed account of important events in the formation and development of the profession. Mainly of interest to professional surveyors and cartographers but also a useful reference for historians.

**253 Thomson, D. W.** *Men and meridians: the history of surveying and mapping in Canada.* Ottawa: R. Duhamel, Queen's Printer, 1966–69. 3 vols.
Relates the many problems encountered by surveyors from the years of earliest settlement. Useful on various aspects of the history of mapmaking and surveying.

**254 Thrower, Norman J. W.** *Original survey and land subdivision: a comparative study of the form and effect of contrasting cadastral surveys.* Chicago: Rand McNally, for the Association of American Geographers, 1966. 160pp. (Association of American Geographers Monograph No. 4).
Concentrates on two areas in western Ohio which Thrower uses to exemplify differing methods used in the subdivision of land.

## Property surveys: bibliography

**255 Eller, Robert C.** *Technical division on property surveys. Bibliography of property surveying literature, compiled by Winfield H. Eldridge.* Washington, D.C.: American Congress on Surveying and Mapping, 1963.
Contains over 1,000 citations.

# Symbols, Ornamentation and Calligraphy

**256 Dainville, François de.** *Le langage des géographes: termes: signes: des cartes anciennes 1500–1800.* Paris: Editions A. et J. Picard & Cie., 1964. 375pp.
An exhaustive exploration of geographical terms with historical outlines of the development of map symbols and map colouring between 1500 and 1800. The illustrations show how mapmakers have developed a specialized cartographic language in the form of both pictorial and abstract symbols to represent environmental features. A unique work that is a valuable reference aid for cartographers, carto-historians, geographers, graphic designers, etc.

**257 George, Wilma.** *Animals & maps.* London: Secker & Warburg, 1969. 235pp.
It has often been said that early mapmakers used drawings of flora and fauna merely as space filling when their knowledge of continental interiors was limited. Dr. George suggests that, on the contrary, cartographers had a sound knowledge of the distribution of animals. The preface by Dr. Helen Wallis is a useful introduction to the history of cartography. Provides an extensive list of references. Of interest to biologists as well as carto-historians and geographers.

**258 Grosjean, Georges** and **Rudolf Kinauer.** *Kartenkunst und Kartentechnik vom Altertum bis zum Barock.* Berne and Stuttgart: Verlag Hallwag, 1970. 144pp.

A superbly illustrated account in colour and black and white of cartographic development, particularly from the artistic angle. Should interest graphic artists and designers as well as carto-historians, etc.

**259 Harvey, P. D. A.** *The history of topographical maps: symbols, pictures and surveys.* London: Thames & Hudson, 1980. 199pp.

Professor Harvey examines the development of topographical maps from the early attempts to communicate information by abstract symbols, through picture maps, to scale maps based on measured survey. An interesting, well-illustrated text and thesis for historians of mapmaking, cartographers, designers, geographers, and map users generally.

**260 Lynam, Edward.** *The mapmaker's art: essays on the history of maps.* London: The Batchworth Press, 1953.

The late Superintendent of the Map Room in the British Museum, in a classic essay ('Period ornament, writing and symbols on maps 1250–1800', pp. 37–54), examines the development of topographical symbols, as well as calligraphy and ornamental features such as the cartouche and border. An interesting and useful reference for cartographers, carto-historians, geographers and graphic designers.

## History of Map Reproduction

Map reproduction is covered to some extent in the major works on the history of mapmaking but there are two important works that look at map reproduction more thoroughly:

**261 Woodward, David** (ed.). *Five centuries of map printing.* Chicago and London: University of Chicago Press, 1975. 177pp.

Based on the third series of Kenneth Nebenzahl Jr. Lectures in the History of Cartography at the Newberry Library in 1972. A highly distinguished body of contributors provide essays on mapmaking and map printing; the evolution of a working relationship; the woodcut technique; copperplate printing; lithography and maps; miscellaneous map printing processes in the nineteenth century; the application of photography to map printing; and the transition to offset lithography. There is an excellent bibliography of ninety-three citations. A unique and valuable reference work for historians, carto-historians, geographers, reprographers, printers, photographers and graphic artists.

**262 Woodward, David.** *The all-American map: wax engraving and its influence on cartography.* London and Chicago: University of Chicago Press, 1977. 168pp.

Widely used in the United States from the mid nineteenth century to the 1960s, wax engraving or cerography exerted a significant influence on the style of American maps. This book relates the history of the craft, largely on the basis of interviews with craftsmen. For carto-historians, geographers, reprographers and designers.

# Dating and Identification of Early Maps

The various carto-bibliographies and checklists such as those of Chubb [194] or Skelton [195] offer valuable assistance in dating and identifying particular maps. So too do the listings of maps of individual counties such as *Printed maps and town plans of Bedfordshire 1576–1900* by Betty Chambers (Bedfordshire Historical Record Society, 1983. 250pp.). There are, however, one or two books especially concerned with the topic, along with some important and helpful papers.

**263 Moreland, Carl** and **David Bannister.** *Antique maps: a collector's handbook.* London and New York: Longman, 1983. 314pp.

This book is divided into three parts; the first two provide the historical background with the principal works of famous cartographers noted and illustrated. The third part of the book is concerned with buying maps and forming a collection.

## Variants in printed maps

**264 Verner, Coolie.** 'The identification and designation of variants in the study of early printed maps'. *Imago Mundi*, **19**: 100–105 (1965).

A useful aid to scholars, historians and collectors in identifying and describing the state of an early printed map.

**265 Verner, Coolie.** 'Carto-bibliographical description: the analysis of variants in maps printed from copperplates'. *The American Cartographer*, **1**(1): 77–87 (1974).

Coolie Verner's paper describes ways of identifying, categorizing and describing the changes made to early printed maps.

## Watermarks

**266 Heawood, Edward.** 'The use of watermarks in dating old maps and documents'. *Geographical Journal*, **63**: 391–412 (1924). Reprinted in Raymond Lister's *How to identify old maps and globes*. London: Bell, 1965.

Watermarks can be a useful aid in the authentication of old documents. Heawood discusses the history of watermarks and details different types used in each country at different times. With monochrome illustrations.

**267 Gravell, Thomas L.** and **G. Miller.** *A catalogue of American watermarks, 1690–1835.* New York: Garland Publishing, 1979. 230pp.

A particularly valuable reference work for those interested in the history of mapmaking in America.

## Special Types of Map

### Celestial cartography: bibliography

**268 Warner, Debora J.** *The sky explored: celestial cartography 1500–1800.* Washington, D.C.: Smithsonian Institution, 1979. 312pp.
A descriptive, illustrated catalogue of celestial maps printed in Europe between 1500 and 1800. For carto-historians, astronomers, cartographers, etc.

### Geological mapping—development

**269 Basset, D. A.** *A source-book of geological, geomorphological and soil maps for Wales and the Welsh Borders (1800–1966).* Cardiff: National Museum of Wales, 1967. 239pp.
Includes a useful summary of the evolution of geological, geomorphological and pedological cartography, with extensive notes on the sources of information. Bibliography pp. 195–211. Reference work for geologists, geographers, soil scientists, geomorphologists, cartographers, librarians and carto-historians.

**270 Boud, R. C.** 'The early development of British geological maps'. *Imago Mundi*, **27**: 73–96 (1975).
A well-illustrated survey of the development of British geological maps from the first proposals by Lister and Aubrey, 1683–91. A good summary for geologists, carto-historians and geographers.

**271 Eyles, V. A.** 'Mineralogical maps as forerunners of modern geological maps'. *The Cartographic Journal*, **9**(2): 133–134 (December 1972).
Traces the development of geological cartography from the eighteenth-century practice of marking mineral deposits on topographical maps to present-day geological mapping showing the distribution of different formations. Generally aimed at cartographers, geology students and historians.

**272 Eyles, V. A.** and **Joan M. Eyles.** 'On the different issues of the first geological map of England and Wales'. *Annals of Science*, **3**: 190–212 (1938).
Discusses the issues of the famous geological map of England and Wales published in 1815 by William Smith, the 'father of English geology'. A useful reference for carto-historians, librarians and map curators as well as geologists.

**273 Ireland, H. A.** 'History of the development of geologic maps'. *Bulletin of the Geological Society of America*, **54**: 1227–1280 (1943).
A general history designed for the professional geologist but with relevance for geographers, carto-historians, map curators and librarians.

**274 North, F. G.** *Geological maps: their history and development, with special reference to Wales.* Cardiff: National Museum of Wales, 1967. 239pp.
Of interest to geologists, historians of thematic mapping, students of Welsh cartography and map curators.

*Bibliography*

**275 Eyles, J. M.** 'William Smith (1769–1839): a bibliography of his published writings, maps and geological sections, printed and lithographed'. *Journal of the Society for the Bibliography of Natural History*, **5:** 87–109 (1969).
An important bibliography of the works of the 'father of English geology' for geologists, historians, cartographers and map curators.

## Guidebook maps

**276 Otness, Harold M.** 'Guidebook maps'. Special Libraries Association, Geography and Map Division, *Bulletin*, **88:** 17–23 (June 1972).
A brief survey of the development of maps in guidebooks. A useful source of information not easily found elsewhere that will be helpful for librarians, geographers, urban scholars and map users generally.

## Medical cartography: maps of health and disease

**277** The cholera epidemic of 1831 brought about the first known mapping of disease in Britain. Several doctors included maps in their reports and writings. Dr. Robert Baker included a map of cholera in Leeds in his Report of the Leeds Board of Health (1833); Dr. Thomas Shapter mapped deaths from cholera in Exeter in his *History of the cholera in Exeter in 1832*; the renowned geographer, Augustus Petermann, prepared a cholera map of the British Isles in 1852 which showed the areas affected in 1831–33; Heinrich Berghaus in his famous *Physikalischer Atlas* introduced maps of the distribution of disease; the second edition of A. K. Johnston's *The physical atlas of natural phenomena* included a world map showing the distribution of disease; Dr. John Snow plotted the deaths from cholera in the Broad Street area of London and showed how these deaths could be traced to people who had drunk from a particular pump. These maps are discussed in the following important paper:

**278 Gilbert, E. W.** 'Pioneer maps of health and disease in England'. *The Geographical Journal*, **CXXIV**(2): 172–183 (June 1958).
This is essential reading for carto- and medical historians, geographers and social scientists, and illustrates all the pioneering maps produced in the middle decades of the nineteenth century.

## Military maps

*Bibliography*

**279 Nebenzahl, Kenneth.** *A bibliography of printed battle plans of the American Revolution, 1775–1795.* Chicago and London: University of Chicago Press, 1975. 159pp.
Nebenzahl records only plans bearing military information and uses a chronological arrangement under theatres of action. An essential reference source for military historians.

*Selected works*

**280 Clark, David S.** *Index to maps of the French and Indian war in books and periodicals illustrating the background of the conflict, British and French military operations in North America, the Cherokee War, the Havana Campaign, and post-war boundaries.* Fayetteville, North Carolina: the author, 1974. 118pp.

**281 Clark, David S.** *Index to maps of the American Revolution in books and periodicals illustrating the Revolutionary War and other events of the period 1763–1789.* Westport, Connecticut: Greenwood Press, 1974. 288pp.

**282 Easton, William W.** 'A history of military mapping; its evolution and use'. Special Libraries Association, Geography and Maps Division, *Bulletin*, **109**: 40–44 (September 1977).
Includes a short bibliography of fifteen items.

**283 Guthorn, Peter J.** *British maps of the American Revolution.* Monmouth Beach, New Jersey: Philip Freneau Press, 1972. 72pp.
This volume has been produced as a companion volume to the same author's *American maps and map makers of the Revolution* and lists around 1,000 maps made by British Army engineers and American loyalists.

**284 Harley, J. B.** with **Barbara Bartz Petchenik** and **L. W. Towner.** *Mapping the American Revolutionary War.* Chicago: University of Chicago Press, for the Hermon Dunlap Smith Center for the History of Cartography at the Newberry Library, 1978. 187pp.
Discusses both contemporary and modern mapping of the war. Includes useful notes and a selective ten-page bibliography.

**285 Marshall, Douglas W.** 'Military maps of the eighteenth century and the Tower of London Drawing Room'. *Imago Mundi*, **32**: 21–44 (1980).
Defines how the Tower of London Drawing Room fitted into a large military network and how its development altered in relation to changes taking place within the Ordnance and the army. Extensive list of references.

**286 Ristow, Walter W.** 'Maps of the American Revolution: a preliminary survey'. *The Canadian Cartographer*, **10**(1): 1–20 (June 1973).
A useful survey of some of the general and specialized cartographic materials of the Revolutionary War period, with particular reference to those in the collections of the Library of Congress's Geography and Map Division. A good reference source for military and carto-historians.

**287 U.S. Library of Congress.** *Maps and charts of North America and the West Indies, 1750–1789.* Washington, D.C.: Library of Congress, 1981. 495pp.
Describes some 2,000 maps, many of which were produced for military reasons: includes maps of land and sea battles, reconnaissance maps, maps showing lines of fire and lines of march, plans of forts, etc. An important reference work of particular value to military historians.

**288 U.S. National Archives.** *Civil War maps in the National Archives.* Washington, D.C.: Government Printing Office, 1964. 127pp.
Lists and describes around 8,000 cartographical items relating to the American Civil War.

**289 Verner, Coolie.** *Maps of the Yorktown campaign 1780–1781; a preliminary checklist of printed and manuscript maps prior to 1800.* London: Map Collectors' Circle, 1965. (Map Collectors' series No. 18.)

# Route maps

*Road-books and road maps: bibliography*

**290 Bonacker, W.** *Bibliographie der Strassenkarten.* Bonn–Bad Godesberg: Kirschbaum, 1973. 242pp.
The most comprehensive bibliography of road maps, listing some 4,600 items.

*Road-books and road maps: lists and catalogues*

**291 Fordham,** *Sir* **H. G.** *The road-books and itineraries of Great Britain, 1570 to 1850.* Cambridge: Cambridge University Press, 1924.
The only existing catalogue of its kind. With a bibliography.

**292 Fordham,** *Sir* **H. G.** *The road-books and itineraries of Ireland, 1647–1850.* Dublin: Falconer, 1923.
A catalogue that serves to supplement Fordham's list of road-books of Great Britain. Together the two volumes form the most useful guide available for librarians and historians, etc.

**293 Nicholson, T. R.** *Wheels on the road: road maps of Britain, 1870–1940.* Norwich: Geo Books, 1983. 101pp.
Gives detailed treatment of the more important publishers and maps. There is a carto-bibliography on pp. 89–101. As well as to carto-historians, of general interest to mapmakers, transport historians, cyclists and motorists.

*Road-books and road maps: selected papers*

**294 Ristow, W. W.** 'American road maps and guides'. *Scientific Monthly,* pp. 397–406 (May 1946).

**295 Ristow, W. W.** 'A half century of oil-company road maps'. *Surveying and Mapping,* **XXIV**(4): 617–637 (December 1964).
Two very useful papers by Ristow on typically American cartographic products. The latter paper is a fascinating discussion of the history of oil-company maps and gives a good background for students of transport and communications as well as for carto-historians and map lovers in general.

*Railway maps: bibliography*

**296 Modelski, Andrew M.** (compiler). *Railroad maps of the United States: a* *selective annotated bibliography of original 19th-century maps in the Geography and Map Division of the Library of Congress.* Washington, D.C.: Library of Congress, 1975. 112pp.

Lists 622 maps that highlight the development of railroad mapping in nineteenth-century America and reflect the changing style and techniques. Good source material for librarians and railway buffs.

## Thematic Mapping

**297 Friis, Herman R.** 'Statistical cartography in the United States prior to 1870 and the role of Joseph C. G. Kennedy and the U.S. Census Office'. *The American Cartographer*, **1**(2): 131–157 (October 1974).

An important paper that describes highlights of the development of statistical cartography in the United States and also includes references to similar progress in Europe. Includes a list of 100 references. Valuable reference material for statisticians, cartographers, economists, historians, social scientists, etc.

**298 Harms, Hans.** *Themen alter Karten.* Oldenburg: Ernst Völker, 1979. 279pp.
Harms examines over 100 maps dating between 1515 and 1804 which were made specially to illustrate particular themes. All are illustrated. There is a four-page bibliography of works relating to thematic maps. A reference for historians, carto-historians and geographers.

**299 Imhof, Eduard.** *Thematische Kartographie.* Berlin: Walter de Gruyter, 1972.
Dr. Imhof contends that thematic maps are as early in origin as maps in general.

**300 Robinson, Arthur H.** *Early thematic mapping in the history of cartography.* Chicago and London: University of Chicago Press, 1982. 288pp.
In this well-illustrated (thirty-nine colour plates and seventy-one monochrome illustrations) and long-awaited volume Professor Robinson shows how, in the late 1600s, mapmakers began to produce maps on specific themes such as vegetation, population and the incidence of disease. By the mid 1800s such maps were fairly common and Robinson examines their effect on the growth of science and mathematics, the introduction of censuses and the epidemics of disease. A major work and the only treatise on the history of thematic maps in any depth. Useful ground for historians, mapmakers, geographers, and students of medicine, the social sciences and demography.

**301 Robinson, Arthur H.** 'The 1837 maps of Henry Dury Harness'. *Geographical Journal*, **121**: 440–450 (1955).
Robinson discusses the important pioneering statistical maps made by Harness for a report to the Irish Railways Commission in 1838. His maps introduce statistical presentation techniques that remain fundamental today. A useful reference for geographers, historians, cartographers, social scientists and those in related disciplines.

# Unconventional Maps and Cartographical Curiosities

**302 Hill, Gillian.** *Cartographical curiosities*. London: British Museum Publications Ltd., for the British Library, 1978. 63pp.

This volume describes eighty-six of the more unusual maps in the Map Library and the Department of Manuscripts which were shown at a special exhibition in April 1978. Exhibits described include maps in board games, jigsaw puzzles and on playing cards; maps of fictitious lands; maps depicting geographical areas as animals or human figures; political cartoon maps and so on. Recommended for librarians and map curators, historians and map enthusiasts generally.

**303 Tooley, R. V.** *Geographical oddities*. London: Map Collectors' Circle, 1963. 22pp. text. (Map Collectors' series No. 1.)

Generally aimed at map collectors.

**304 Tooley, R. V.** *Leo Belgicus: a list of variants*. London. Map Collectors' Circle. 16pp. text. (Map Collectors' series No. 7.)

Tooley discusses the numerous examples of the map of the Netherlands drawn in the form of a lion which have proved the most familiar of all animal maps.

# Tapestry Maps

**305 Bedford, W. K. R.** 'The Weston tapestry maps'. *Geographical Journal*, **9**: 210–215 (1897).

Describes the five magnificent tapestry maps dated *c.* 1570–88 which are somewhat similar in design to Saxton's county maps. Of interest to carto-historians, needlewomen and designers.

# Globes

## Bibliography

**306 Bonacker, Wilhelm.** *Das Schrifttum zur Globenkunde*. Leiden: E. J. Brill, 1960. 58pp.

Lists 660 references concerning the history of globes. With author index.

## Lists and catalogues

**307 Duprat, Gabrielle.** *Liste des globes terrestres et célestes anciens—antérieurs à 1850—conservés dans les collections publiques de France*. Paris: Centre National de la Recherche Scientifique, 1970. 42pp.

Lists 231 globes in French public collections.

**308 Fauser, Alois** and **Traudl Seifert.** *Altere Erd- und Himmelsgloben in Bayern. Im Auftrag der Bayerischen Staatsbibliothek*. Stuttgart: Schuler Verlag, 1964. 184pp.

Lists 246 globes and discusses early examples preserved in Bavarian libraries. A well-illustrated reference.

**309 Fiorini, Matteo.** *Sfere terrestri e celesti di autore italiano oppure fatte o conservate in Italia.* Rome: La Società Geografica Italiana, 1899. 502pp.
Lists globes by Italian makers or globes preserved in Italy and made before 1800.

**310 Grötzsch, H.** (ed.). *Die ersten Forschungsergebnisse der Globusinventarisierung in der Deutschen Demokratischen Republik.* Berlin: VEB Deutscher Verlag der Wissenschaft, 1962. 202pp.
Lists globes preserved in East Germany.

**311 Krogt, Peter van der.** *Old globes in the Netherlands.* Utrecht: HES Publishers, 1984. 294pp.
A union catalogue of pre-1850 globes in the Netherlands, with 340 entries.

**312 Luzio, L.** *Catalogo dei globi antichi conservati in Italia.* Vol. 1, *I globi Blaviani.* Florence: Istituto e Museo di Storia della Scienza, Biblioteca, 1957. 54pp.
Devoted to the seventeenth-century globes of the Blaeu establishment.

**313 Yonge, Ena L.** *A catalogue of early globes, made prior to 1850 and conserved in the United States: a preliminary listing.* New York: American Geographical Society, 1968. 118pp. (Library Series No. 6.)
Lists some 400 globes, armillary spheres and orreries and provides brief biographical notes on the globemakers mentioned. One of the best sources of information on globes in the English language and a valuable reference for librarians, curators of map collections and carto-historians.

## Histories of globes and their manufacture

**314 Fauser, Alois.** *Die Welt in Händen. Kurze Kulturgeschichte des Globus.* Stuttgart: Schuler Verlag, 1967. 184pp.
German text. A well-illustrated history of early globes and their makers.

**315 Muris, Oswald** and **Gert Saarmann.** *Der Globus im Wandel der Zeiten: eine Geschichte der Globen.* Berlin: Columbus Verlag, 1961. 287pp.
German text. A useful history with over 120 globes cited.

**316 Stevenson, E. L.** *Terrestrial and celestial globes; their history and construction, including a consideration of their value as aids in the study of geography and astronomy.* New Haven, Connecticut: Yale University Press, for the Hispanic Society of America, 1921. 2 vols. (Hispanic Society of America Publication No. 86.)
A comprehensive work that includes a bibliographical listing in the second volume. Valuable for astronomers as well as carto-historians and geographers.

## Identification and dating

**317 Baynes-Cope, A. D.** 'The investigation of a group of globes'. *Imago Mundi*, **33**: 9–20 (1981).
Includes useful information on the manufacture and dating of globes.

*Periodical*

**318** *Der Globusfreund.* Vienna: Coronelli-Weltbund der Globusfreunde.
An annual publication entirely devoted to the study of modern and historical
globes.

# Map Collecting

**319 Skelton, R. A.** *Maps: a historical survey of their study and collecting.* Chicago and
  London: University of Chicago Press, 1972. 138pp.
This scholarly work provides the first systematic history of map collecting and
shows how the history of cartography developed as a field of study. Of great
importance to carto-historians, antiquarians, librarians and curators of maps,
and full of interest for the general collector.

### General works specifically designed for collectors

**320 Baynton-Williams, R.** *Investing in maps.* London: Barrie & Rockliff, The
  Cressett Press, 1969. 160pp.
A well-illustrated (colour and black and white) guide to map collecting for
beginners. The author discusses the range and variety of maps available and
gives some indication of relative values. Written by a London dealer.

**321 Hodgkiss, A. G.** *Discovering antique maps.* 5th edn. Princes Risborough: Shire
  Publications, 1983. 72pp.
A brief introduction to early maps that looks at their history and characteristics
and aims to demonstrate how much more they have to offer than mere
decoration.

**322 Lister, Raymond.** *How to identify old maps and globes, with a list of cartogra-*
  *phers, engravers, publishers and printers concerned with printed maps and globes from* c.
  *1500 to* c. *1850.* London: G. Bell & Sons Ltd.; Hamden, Connecticut: Archon
  Books, 1965. 265pp.
Includes well-illustrated chapters on the history of maps and charts; celestial
maps; map reproduction; decoration; conventional signs. The appendix includes
Heawood's valuable essay on the use of watermarks in dating old maps.

**323 Lister, Raymond.** *Antique maps and their cartographers.* London: G. Bell &
  Sons, 1970. 128pp.
A short history of mapmaking designed primarily for amateur collectors.

**324 Moreland, Carl** and **David Bannister.** *Antique maps: A collector's handbook.*
  London and New York: Longman, 1983. 314pp.
Written by a map collector and a map dealer, this book attempts to fulfil the
need for an exhaustive handbook that will provide information for map
collectors, and particularly those new to collecting.

**325 Radford, P. J.** *Antique maps.* London: Garnstone Press, 1971. 72pp.
A brief introduction by a well-known dealer. Lists over 100 important mapmakers and gives short details of their output.

*Journals on the history of cartography and map collecting*

**326** *Imago Mundi.* E. M. J. Campbell (ed.). c/o King's College, London. 1935–. Annual.
A scholarly compilation of papers, reviews, news of current events, etc. which provides serious students of early mapmaking with valuable reference material. Annual issues with irregular supplementary issues devoted to specific topics.

**327** *The Map Collector.* Valerie G. Scott (ed.). P.O. Box 53, Tring, Hertfordshire HP23 5BH, England. Quarterly.
A quarterly journal published in March, June, September and December that contains scholarly articles on the history of cartography as well as reviews, news of current events, an auction section, an important directory of dealers and a collectors' marketplace. Superbly illustrated.

**328** *IMCOS Journal.* The quarterly publication of the International Map Collectors' Society. Editor: Yasha Beresiner, 1A Camden Walk, Islington Green, London N1 8DY, England.
Contains articles on early maps; editorial news and views; current events; reports on meetings, etc.

## Facsimile Reproductions of Early Maps

**329 Hodgkiss, A. G.** 'Facsimiles of early maps'. *SUC Bulletin*, **12**(2): 1–12 (1978).
Includes an outline history of facsimile production and a discussion of the wide range of facsimile publishing today. Includes also a useful chronological list of facsimiles of important maps and atlases as well as a list of publishers and distributors of early map facsimiles. A useful reference in a field rather neglected by authors. For collectors, historians, geographers and students of cartography.

**330 Ristow, Walter W.** and **Mary E. Graziani.** *Facsimiles of rare historical maps: a list of reproductions for sale by various publishers and distributors.* 3rd edn. Washington, D.C.: Government Printing Office, 1968. 12pp.
A valuable list with an index of available facsimiles. Recommended for historians as a guide to some basic source materials and to geographers, cartographers, illustrators and collectors.

**331 Ristow, Walter W.** 'New maps from old: trends in cartographic facsimile publishing'. *The Canadian Cartographer*, **5**(1): 1–17 (June 1968).
Dr. Ristow traces the development of the techniques of facsimile publishing from the mid nineteenth century to the present day. A valuable reference guide for researchers into cartographic history that should also appeal to reprographers, collectors and map lovers in general.

**332 Modelski, Andrew M.** 'Introductory essays and commentaries for the study of the history of cartography from selected facsimile atlases in the Library of Congress'. Special Libraries Association, Geography and Map Division, *Bulletin*, No. 116: 43–51 (June 1979).

A useful bibliography of the many important essays used to introduce modern facsimile reproductions of early atlases. Good reference material for historians, carto-historians, librarians, map curators and others.

**333 Noe, Barbara.** *Facsimiles of maps and atlases. A list of reproductions for sale by various publishers and distributors.* 4th edn. With an introduction by John A. Wolter. Washington, D.C.: Superintendent of Documents, Government Printing Office, 1978.

This is a revised edition of Ristow's and Graziani's work [330]. Around 500 facsimile maps and atlases from 145 publishers are described.

PART III

Annotated Bibliography of Reference
Sources: Contemporary Cartography

# Information Sources on Contemporary Cartography

## General Sources

### Use of bibliographies

**334 Drazniowsky, Roman.** 'Bibliographies as tools for map acquisition and map compilation'. *The Cartographer*, **3**(2): 138–144 (December 1966).
Drazniowsky examines a variety of information sources relevant to: (a) maps, atlases and globes; (b) general cartography—the nature of cartography, its history and methods; and (c) bibliographic aids and gazetteers. Includes a selective bibliography of fifty-two entries.

### Carto-bibliographies: general

**335 Library of Congress.** *The bibliography of cartography*. Washington, D.C.: The Library of Congress, Geography and Map Division, 1973. 5 vols. (Distributed by G. K. Hall, 70 Lincoln Street, Boston, Massachusetts 02111.)
As the most comprehensive work of its kind, this is a standard reference source for libraries, higher educational establishments and official agencies.

**336 Zögner, Lothar** (ed.). *Bibliographia Cartographica*. Staatsbibliothek Preussischer Kulturbesitz with the Deutsche Gesellschaft für Kartographie. (Available from K. G. Saur Verlag KG, Pf 71 1009, D-8000 München 71, Federal Republic of Germany.)
The annual issues record all kinds of cartographical literature including articles from two hundred periodicals. A valuable reference for librarians and curators of collections. It replaces *Bibliotheca Cartographica*, which was issued annually from 1957 to 1972 by the Institut für Landeskunde in the Bundesforschungsanstalt für Landeskunde und Raumordnung with the Deutsche Gesellschaft für Kartographie.

**337** *Bibliographie Cartographique Internationale.* Paris: Centre National de la Recherche Scientifique, in collaboration with UNESCO and the International Geographical Union, 1949–76.
This annual map bibliography was published in Paris and provided a list of current maps published in the world. Its information was contributed by thirty countries. Publication has now ceased.

**338** *Bibliographic guide to maps and atlases, 1979.* Boston: G. K. Hall, 1980. 633pp.
A guide not only to cartographic products but also to current literature on cartography.

**339** *World Cartography.* New York: UN Dept. of Economic and Social Affairs, 1976. Volume 14. 108pp.
Devoted to the current status of world topographic mapping. An essential reference for map curators and librarians.

**340 American Geographical Society.** *Index to maps in books and periodicals.* New York: Map Department, American Geographical Society, 1967. 10 vols.
Complements carto-bibliographies that record separately issued maps.

### Carto-bibliographies: national

**341 Andriot, John L.** *Guide to U.S. government maps.* McLean, Virginia: 1975. Documents Index. Box 195. 2 vols.
The first volume of 309 pages is a reprint of the placenames appearing in the index of the *National Atlas of the United States of America* (1970) and the second volume lists almost 5,000 geological and hydrological maps published by the U.S. Geological Survey between 1879 and 1974. A useful reference for librarians, map curators, and all users of U.S. maps.

**342** *Australian maps, 1961–1973.* Canberra: National Library of Australia, 1975. 234pp.
*Australian maps, 1974–.* Canberra: National Library of Australia, 1975.
These two volumes cover maps representing areas of Australia or by Australian authors. They include a useful list of Australian map publishers. Subscribers to *Australian maps* also receive a copy of *Overseas map acquisitions, 1975–*, which describes the foreign acquisitions made by the Australian National Library.

**343** *Bibliografie van in Nederland Verschienen Kaarten, 1975.* The Hague: Koninklijke Bibliotheek, 1980. 224pp.
One of an annual series that lists all recently published maps in the Netherlands.

**344 Larsgaard, M. L.** *Topographic mapping of the Americas, Australia and New Zealand.* Littleton, Colorado: Libraries Unlimited, 1984. 180pp.
A valuable modern survey of topographic mapping, with a very full bibliography.

**345 Nicholson, N. L.** and **L. M. Sebert.** *The maps of Canada: a guide to official*

*Canadian maps, charts, atlases and gazetteers.* Folkestone: Wm. Dawson & Sons; Hamden, Connecticut: Archon Books, 1981. 251pp.
Deals with the current state of Canadian mapping and also with its history. A useful source for users of Canadian maps or for those teaching others their use. Includes bibliographies, glossary, maps and tables.

**346 Thompson, Morris Mordecai.** *Maps for America. Cartographic products of the U.S. Geological Survey and others.* Reston, Virginia: U.S. Geological Survey, 1979. 265pp. (Available from the Superintendent of Documents, Government Printing Office, Washington, D.C. 20402.)
An important reference work for librarians and map curators, geographers, geologists, carto-historians and all users of United States official maps. Illustrated with map sections.

**347 U.S. National Archives.** *Guide to cartographic records in the National Archives.* Washington, D.C.: U.S. National Archives Publications 71–16. (Available from the Superintendent of Documents, Government Printing Office, Washington, D.C. 20402.)
An important aid in using the 1.6 million maps and 2.25 million aerial photographs in the National Archives.

## Selected works

**348 Birch, T. W.** *Maps, topographical and statistical.* 2nd edn. Oxford: Clarendon Press, 1964. 240pp.
A general introduction to mapmaking suitable for first-year courses in cartography, geography or earth sciences.

**349 Centre Georges Pompidou.** *Cartes et figures de la Terre.* Paris: 1980. 480pp.
The sumptuous catalogue of an exhibition held in the Pompidou Centre. The text, in French, covers all types of map. For laymen and professionals.

**350 Cuff, David J.** and **Mark T. Mattson.** *Thematic maps: their design and production.* New York and London: Methuen, 1982. 169pp.
An up-to-date introduction to the principles and practices of thematic mapmaking. Ideally suited to first-year courses in geography or cartography and a valuable reference work for anyone called upon to prepare maps for illustrative purposes. Clearly written and well illustrated.

**351 Cuenin, R.** *Cartographie générale.* Paris: Editions Eyrolles, 1973. Vol. 1, 324pp.; vol. 2, 208pp.
The first volume covers general concepts and principles; the second deals with reproduction and printing techniques, compilation and editing.

**352 Heissler, V.** and **G. Hake.** *Kartographie.* Berlin: Walter de Gruyter & Co., 1970. Vol. I, 233pp.; vol. II (by G. Hake), 202pp.
Technical volumes designed for teachers as well as students of mapmaking.

**353 Hodgkiss, A. G.** *Understanding maps: a systematic history of their use and development.* Folkestone: Wm. Dawson & Sons, 1981. 208pp.
Includes not only historical material but also chapters on various kinds of modern map. For the general reader, collectors, students of geography and cartography, map users in general.

**354 Koeman, Cornelis** *et al. Basic cartography for students and technicians.* International Cartographic Association/Association Cartographique Internationale (available from R. Muller, P.O. Box 9016, NL-1006 AA Amsterdam). Vol. 1, 1984. 206pp. Vol. 2 (in preparation).
A best-selling and most important text on cartography that is available in several different language editions.

**355 Lawrence, G. R. P.** *Cartographic methods.* 2nd edn. London: Methuen, 1979. 153pp.
A useful introduction, largely concentrating on statistical mapmaking, designed for first-year courses in geography at university level.

**356 Muehrcke, Phillip C.** *Map use: reading, analysis and interpretation.* Madison, Wisconsin: JP Publications, 1978. 469pp.
Written strictly for 'the person who wants to use maps'. Muehrcke divides his text into four parts: Map Reading; Map Analysis; Map Interpretation; Orientation. Appendices cover Sources of Maps; Remote Sensing; Map Projections; Enlarging and Reducing; Map Care. A lively, stimulating work that should appeal to all with an interest in maps but particularly to students of cartography and geography.

**357 Raisz, Erwin.** *General cartography.* 2nd edn. New York: McGraw-Hill, 1948. 270pp.
Designed to help the student understand the language of maps; to enable him to illustrate his own papers and to give him a foundation if he chooses to take up cartography as a profession. Suitable for first-year courses in departments of higher education, for sixth forms or for the general reader. Inevitably a little dated.

**358 Rhind, David** (ed.). 'Contemporary cartography'. *Transactions* of the Institute of British Geographers, New Series, **2**(1) (1977).
A. H. Robinson, J. L. Morrison and P. C. Muehrcke look at the changes that have taken place since 1950 and try to assess developments up to the year 2000. Christopher Board and R. M. Taylor study perception and maps and analyse the human factors in map design and interpretation. For geographers and professional mapmakers.

**359 Robinson, A. H., Randall Sale** and **Joel Morrison.** *Elements of cartography.* 4th edn. New York, Santa Barbara, Chichester, Brisbane and Toronto: John Wiley, 1978. 448pp.
The most comprehensive work of its kind. Well illustrated. The most appropriate

manual for first-year and more advanced courses in cartography and useful for courses in geography and as a reference for the professional cartographer.

**360 Robinson, A. H.** and **Barbara Bartz Petchenik.** *The nature of maps: essays toward understanding maps and mapping.* Chicago and London: University of Chicago Press, 1976. 138pp.
A philosophical work that attempts to correlate all the elements of cartography, from the cartographer's perception of what is to be mapped, through the cartographic execution process, to the final perception of what is being communicated by the map. Suitable for professional cartographers and geographers.

**361 Salichtchev, Konstantin Alekseevich.** *Osnovy kartovedeniia: istoriia kartografii i kartograficheskie istochniki* [Principles of cartography: historical cartography and cartographic materials]. 3rd revised edn. Moscow: Geodezizdat, 1959–62. 2 vols.
An important general work in Russian that includes a section on the history of cartography.

## Visual aids

**362** *Introduction to cartography: an audio-visual series.* Sydney: Department of Geography, University of Sydney.
An audio-visual presentation that consists of three sets of slides, providing over 130 illustrations of various aspects of mapmaking including a brief examination of the history of cartography. The emphasis is on Australia but the material would be useful for teachers of cartographic courses elsewhere.

## Glossaries of technical terms

**363** A list of national and international glossaries is to be found on pp. 10 and 11 of *Modern maps and atlases* [576].
The following represents a useful selection.

**364** *Dictionary of mapping, charting terms.* Map Division IV, Mail Stop 5096, Madison, Wisconsin 53705. 274pp.

**365** *Glossary of mapping, charting and geodetic terms.* 3rd edn. Washington, D.C.: United States Army Topographic Command, Department of Defense, 1973. 281pp.

**366** *Glossary of technical terms in cartography.* London: British National Committee for Geography, The Royal Society, 1966. 84pp.
Lists technical terms, followed by appendices on projections, isograms, paper sizes. Includes a list of references.

**367** *Multilingual dictionary of technical terms in cartography.* Wiesbaden: Published for the International Cartographic Association by Franz Steiner, 1973. 573pp.

Contains approximately 1,200 terms in five languages, and key words in a further nine languages. An important reference work for map curators and all who have to use foreign maps.

**368 Rosenberg, P., Kent Ericson** and **Gerhardt C. Rowe.** *Digital mapping glossary.* Morristown, New Jersey: Keuffel & Esser Co., 1974. 61pp.
Designed for photogrammetrists and computer scientists engaged in developing and using semi- and fully automatic digital mapping systems.

## Journals

**369** *Acquisitions.* Ottawa: Department of Energy, Mines and Resources, Departmental Map Library, 1967–. Eight times yearly.

**370** *American Cartographer.* Falls Church, Virginia: American Congress on Surveying and Mapping, 1974–. Twice yearly.

**371** *Bibliographie Cartographique Internationale.* Paris: Comité Nationale Français de Géographie of the International Geographic Union, 1936–76. Annual.
Provided international listings of maps and atlases. A good reference aid for curators and librarians.

**372** *Boletín de Información.* Madrid: Servicio Geográfico del Ejército (Geographical Department of the Army). Quarterly.

**373** *Bolletino della Associazione Italiana di Cartografia.* Florence: Associazione Italiana di Cartografia, 1964–. Three times yearly.

**374** *Bulletin.* Ottawa: Association of Canadian Map Librarians, 1967–. Three times yearly.

**375** *Bulletin.* Washington, D.C.: Special Libraries Association, Geography and Map Division, 1947–. Quarterly.
Undoubtedly one of the most useful sources of information about new cartographic material. Includes articles and news items aimed specifically at map curators but important also for cartographers, geographers, librarians and all map users.

**376** *Bulletin du Comité Français de Cartographie.* Paris: 1959–.
Irregular publication containing articles and reviews.

**377** *SUC Bulletin.* Reading: Society of University Cartographers, 1965–. Twice yearly.
Contains articles, news items and reviews aimed primarily at cartographers in higher education but interesting and useful for anyone concerned with map use or production. The extensive review section is probably the best of its kind.

**378** *Cartactual.* Budapest: Hungarian State Institute for Geodesy and Cartography, 1965–. Six times yearly.

An important information service that provides topical information in map form on boundary changes, new names, new industries, transport networks, etc.

**379** *Cartinform.* Budapest: Hungarian State Institute for Geodesy and Cartography, 1971–.
A serialized bibliography of new cartographic material.

**380** *Cartographica.* Formerly known as *The Cartographer* and *The Canadian Cartographer.* Toronto: York University, 1964–. Twice yearly.
An important journal containing papers, articles and reviews on the history of cartography and contemporary mapmaking.

**381** *The Cartographic Journal.* British Cartographic Society, 1964–. Twice yearly.
The official journal of the British Cartographic Society. Contains technical articles, articles on various aspects of the history of cartography, reviews, news items and lists of recent publications.

**382** *Cartography.* Canberra: Australian Institute of Cartographers, 1954–. Twice yearly.
Carries articles, reviews, news items mainly aimed at professional cartographers.

**383** *Geodeticky a Kartograficky Obzor* [Geodetic and Cartographic Review]. Prague: Czech and Slovak Directorate of Geodesy and Cartography, 1955–. Monthly.

**384** *Geodeziia i Kartografiia* [Geodesy and Cartography]. Central Directorate of Geodesy and Cartography of the Council of Ministers of the U.S.S.R., 1925–. Monthly.

**385** *Geodezja i Kartografia* [Geography and Cartography]. Warsaw: Geodetic Committee of the Polish Academy of Sciences. Quarterly.

**386** *Globe.* Clayton, Victoria: Australian Map Circle, 1974–. Annual.
Contains articles concerning maps and atlases, mainly for curators.

**387** *Globen.* Stockholm: Generalstabens Litografiska Anstalt (GLA), 1922–. Quarterly.
Contains short articles on maps, atlases and general cartography.

**388** *Information Bulletin of the Western Association of Map Libraries.* Santa Cruz, California: 1969–. Three times yearly.
Carries articles and reviews with emphasis on western North America.

**389** *International Yearbook of Cartography.* London: George Philip & Son, 1961–. Annual.
Hardback publication with material on all aspects of mapmaking. Some papers are in their original language with an English summary.

**390** *Kartographische    Nachrichten.*    Gütersloh:    Deutsche    Gesellschaft    für Kartographie, 1951–. Six times yearly.
Contains articles and reviews, in German.

**391** *Kartografisch Tijdschrift.* Rijswijk: Nederlandse Vereniging voor Kartografie, 1975–. Quarterly.
Carries articles, reviews, lists of publications, news items.

**392** *Map.* Tokyo: Association of Japanese Cartographers, 1963–. Quarterly.

**393** *Nachrichten aus dem Karten- und Vermessungswesen.* Frankfurt am Main: Institut für Angewandte Geodesie, 1951–. Irregular.
Includes lists of new maps.

**394** *New Geographical Literature and Maps.* London: Royal Geographical Society, 1951–1980. Twice yearly. Now discontinued.
Included a very useful section on recently published maps and atlases.

**395** *New Zealand Cartographic Journal.* Wellington: New Zealand Cartographic Society, 1971–. Twice yearly.

**396** *Polski Przeglad Kartograficzny.* Warsaw: Polskie Towarzystwo Geograficzne, 1969–. Quarterly.
Polish text with English and Russian summaries.

**397** *Revista Brasileira de Cartografia* [Brazilian Journal of Cartography]. Rio de Janeiro: Instituto Brasileiro de Geografia. Twice yearly.

**398** *Revista Cartografica* [Cartographic Journal]. Buenos Aires: Military Geographic Institute, 1951–.

**399** *Survey Review* (formerly *Empire Survey Review*). London: Commonwealth Association of Surveying and Land Economy, 1931–. Quarterly. (Available from C. F. Hodgson & Son Ltd., Unit 4, Central Trading Estate, Staines TW18 4UR, England.)

**400** *Surveying and Mapping.* Washington, D.C.: American Congress on Surveying and Mapping, 1941–. Quarterly.
Carries articles and reviews with an emphasis on surveying.

**401** *World Cartography.* New York: United Nations, 1951–. Irregular.
Includes lists of maps and analyses the state of world mapping.

### Newsletters

**402** *base line* [sic]. Newsletter of the American Library Association's Map and Geography Round Table (MAGERT). 6 issues per annum. (Available on subscription from J. Walsh, *base line* Subscription Manager, Coe Library,

University of Wyoming, University Station, Box 3334, Laramie, Wyoming 82071.)
Provides current information on new publications in any format which relate to maps, cartography and geography, and gives news of conferences, etc.

**403** *Mapline.* Quarterly newsletter published by the Hermon Dunlap Smith Center for the History of Cartography at the Newberry Library. Correspondence to The Editor, The Newberry Library, 60 West Walton Street, Chicago, Illinois 60610.
Carries articles, reviews, items of news about forthcoming conferences, exhibitions, etc. Has a calendar of events.

**404** *Newsletter.* Washington, D.C.: Association of American Geographers. Ten times per year.
Although not solely concerned with cartography it includes reviews of atlases, etc.

**405** *Newsletter.* Society of University Cartographers. Editor: S. Chilton, School of Geography and Planning, Middlesex Polytechnic, Queensway, Enfield, Middlesex, England. Twice per year.
Circulated to members of the society only.

**406** *Newsletter.* British Cartographic Society. Editor: David Fairbairn, Department of Surveying, Old Brewery Building, University of Newcastle, Newcastle upon Tyne NE1 7RU, England. Twice per year.
For members only. As well as general news items, contains an interesting series of profiles describing cartographic organizations.

**407** *Newsletter of the National Cartographic Information Center.* Quarterly publication from the National Cartographic Information Center, U.S. Geological Survey, Mailstop 507, Reston, Virginia 22002.

**408** *Newsletter of the International Cartographic Association.* The sales agent for ICA publications is Rudolf Muller International Booksellers, P.O. Box 9016, NL-1006 AA Amsterdam, The Netherlands.

**409** *Sheetlines.* Newsletter of the Charles Close Society. Available from Alan Godfrey, 57–58 Spoor Street, Dunstan, Gateshead, England.

## Information services

**410 National Cartographic Information Center.** Communications to: Office of the Chief, National Cartographic Information Center, Reston, Virginia 22092.
The National Cartographic Information Center (NCIC) operates a national information service for United States cartographic data. NCIC organizes and distributes descriptions of maps and charts, aerial and satellite photographs, satellite imagery, map data in digital form and geodetic control information. In

addition to the main centre at Reston there are now four regional centres: Menlo Park, California; Denver, Colorado; Rolla, Missouri; and Reston, Virginia.

**411 North American Cartographic Information Society**: c/o Christine Reinhard, 143 Science Hall, University of Wisconsin, Madison, Wisconsin 53706.
The inaugural meeting of this society was held at Milwaukee on 2–4 October 1980. Its purpose and objectives are to promote interest in cartography; to improve understanding of cartographic materials; to promote communication; to improve automated bibliographic control of cartographic materials; and to improve access to cartographic information.

**412 Cartographic Users Advisory Council**: communications to Charles A. Seavey, CUAC, GPMD–General Library, University of New Mexico, Albuquerque, New Mexico 87131.
This body represents four professional map library groups in the United States with the aim of improving communication and distribution.

## Information Gathering

### Field survey

**413 Pugh, J. C.** *Surveying for field scientists*. London: Methuen, 1975. 230pp.
Surveying is dealt with in numerous specialized textbooks; this one is designed as an introduction for geographers, geologists, ecologists and other field scientists who need to map spatial distributions in the field.

### Geodetic survey

**414 Gossett, F. R.** *Manual of geodetic triangulation*. Rockville, Maryland: U.S. Coast and Geodetic Survey, 1971. 302pp.
Describes the methods of triangulation used by the Coast and Geodetic Survey. Useful for students of cartography and surveying as well as field scientists.

**415 Rappleye, Howard S.** *Manual of geodetic leveling*. Rockville, Maryland: U.S. Coast and Geodetic Survey, 1948. 103pp.
Designed to aid field parti.s by illustrating the methods of levelling used in the Coast and Geodetic Survey.

### Urban survey

**416 Blachut, T. J., Adam Chrzanowski** and **J. H. Saastamoinen.** *Urban surveying and mapping*. New York: Springer, 1979. 372pp.
Useful both as a reference work and textbook, this volume covers cartography, surveying, geodesy and photogrammetry. It includes chapters on suitable map projections for urban areas, photogrammetry in urban areas, and an introduc-

tion to the usefulness of computer-based data banks. For students of cartography, local authority surveyors, urban geographers.

## Surveying instruments

**417 Deumlich, Fritz.** *Surveying instruments.* 7th edn. Berlin: Walter de Gruyter & Co., 1982. 316pp.
An authoritative reference work that describes a wide range of instruments, including products from Western and Eastern Europe, Soviet Russia and Japan. It should be in every survey library and is a primary reference work for survey students.

## Survey: research directory

**418** *Directory of research and development in the fields of land survey, geodesy, photogrammetry and hydrographic surveying.* London: published jointly by the Royal Institution of Chartered Surveyors and the Photogrammetric Society, 1982.

## Survey: journals

**419** *Survey Review* (formerly *Empire Survey Review*). London: Commonwealth Association of Surveying and Land Economy, 1931–. Quarterly. Annual subscription to C. F. Hodgson & Son Ltd., Unit 4, Central Trading Estate, Staines, Middlesex TW18 4UR, England.

**420** *Surveying and Mapping.* 210 Little Falls St., Falls Church, Virginia 22046: American Congress on Surveying and Mapping. Quarterly.

# Remote Sensing

Remote sensing is a very wide-ranging topic that would require a whole volume to itself. The following references should help in giving a basic introduction.

### Bibliography

**421 Bryan, M. Leonard.** 'Annotated bibliography of radar remote sensing as used in the geosciences: Part I'. Geography and Map Division, Special Libraries Association, *Bulletin*, **95**: 48–60 (March 1974). Parts II, III, *Bulletin*, **96**, 21–40 (June 1974). Part IV, *Bulletin*, **97**: 16–32 (September 1974).

### Guide to information sources

**422 Bryan, M. Leonard.** *Remote sensing of earth resources: a guide to information sources.* Detroit: Gale Research, 1979. 188pp.
This is the first volume of a Geography and Travel Information Guide Series. Bryan lists 378 citations and describes the scope of each work. Covers general literature, symposia proceedings, manuals, catalogues of remote sensing data, maps derived from remote sensing data, bibliographies, journals, workshops and courses.

*General works*

**423 Barret, G. C.** and **L. F. Curtis.** *Introduction to environmental remote sensing.* London: Chapman & Hall, 1976. 336pp.
Written by, and largely for, geographers. Introductory level.

**424 Burnside, C. D.** *Mapping from aerial photographs.* London: Granada, 1979. 304pp.
A useful practical text.

**425 Colwell, R. N.** *Manual of remote sensing.* 2nd edn. Falls Church, Virginia: American Society of Photogrammetry, 1983. 2 vols. 2,724pp.

**426 Rudd, Robert D.** *Remote sensing: a better view.* North Scituate, Massachusetts: Duxbury Press, 1974. 135pp.
Written primarily for laymen but also provides a useful introduction for geographers, mapmakers and earth scientists requiring a well-illustrated basic text. Explains important developments and emphasizes their significance in resource management. Each chapter has a bibliography.

**427 Slama, C. C.** (ed.). *Manual of photogrammetry.* 4th edn. Falls Church, Virginia: American Society of Photogrammetry, 1980. 1,056pp.
Despite some bias towards the American scene, this is a worthwhile volume for libraries in educational establishments everywhere. Includes a history of remote sensing as well as chapters on all other facets.

**428 Townshend, J. R. G.** (ed.). *Terrain analysis and remote sensing.* London: Allen & Unwin, 1981. 232pp.
One of a series of books designed for scientists working in a specialized field. Generally aimed at those without much previous background in either terrain analysis or remote sensing.

**429 White, Leslie Paul.** *Aerial photography and remote sensing for soil survey.* Oxford and New York: Oxford University Press, 1977. 104pp.
A very useful introduction to remote sensing and its specific application to soil survey which should interest other scientists as well as pedologists.

## Specialized topics

**430 Kroeck, Dick.** *Everyone's space handbook: a photo imagery source manual.* Pilot Rock, Oregon: Arcata, 1976. 175pp.
Identifies the agencies, organizations, etc. that collect, maintain and distribute space data. Includes a short history of aerial photography and space data. A useful reference source for all who use any form of space imagery.

**431 Krumpe, Paul.** *The world remote sensing bibliographic index: a comprehensive geographic index bibliography to remote sensing site investigations of natural and agricultural resources throughout the world.* Fairfax, Virginia: Tensor Industries, 1976. 619pp.

Includes some 4,000 citations by area and source. The area listings are subdivided by discipline. Of prime importance to all earth scientists.

### Journals

**432** *The International Journal of Remote Sensing.* Published for The Remote Sensing Society. Subscription details from the publisher, Taylor & Francis, Rankine Road, Basingstoke RG24 0PR, England.

**433** *The Photogrammetric Record.* The Photogrammetry Society. Two issues per annum. Available from Subscription Department, Department of Photogrammetry and Surveying, University College, Gower Street, London WC1E 6BT, England.

**434** *South African Journal of Photogrammetry, Remote Sensing and Cartography* (formerly *The South African Journal of Photogrammetry*). Published by the South African Society for Photogrammetry, Remote Sensing and Cartography, PO Box 69, Newlands 7725, South Africa.

## Sources of Statistics

Thematic or special-purpose maps can be divided into two main categories, qualitative and quantitative. In the first-named category different types of phenomena are merely distinguished in kind without any attempt being made to assess quantity, intensity or relative size. Themes that can be explored qualitatively include geology, land use, vegetation, religious or language groups and so on, the different types being distinguished by the use of colour or shading patterns. When, however, the mapmaker is concerned with a topic such as population distribution or density he will express the data numerically and produce a map that might show the relative population of cities or the population density per square kilometre. Such maps cannot be produced unless the appropriate data are available. It has been said, however, that anything, whether it be a tangible feature such as a factory, or intangible such as climate, can be mapped providing there are the necessary data.

Items [435]–[452] provide the potential mapmaker with international and national sources of all kinds of statistical data for use in the preparation of many types of map. Items [453]–[459] are concerned with statistical applications to cartography, with data handling and manipulation, and the visual presentation of statistics in the form of maps and diagrams.

## Guides to statistics: general, international

**435 Ball, J.** (ed.). *Foreign statistical documents; a bibliography of general, international trade and agricultural statistics, including holdings of the Stanford University Libraries.* Compiled by R. Gardella. Stanford, California: Hoover Institution on War, Revolution and Peace, Stanford University, 1967. 173pp.
Approximately 1,700 entries on general, trade and agricultural statistics.

**436 Brewer, J. Gordon.** 'Sources of statistics'. *The literature of geography: a guide to its organization and use,* Chapter 6, pp. 80–91. London: Clive Bingley; Hamden, Connecticut: Linnet Books, 1973.
Discusses over 100 sources. The arrangement is: Problems in the use of statistics; Bibliographies; Libraries; International sources; British sources; American sources.

**437 Cormier, Reine.** *Les sources des statistiques actuelles: guide de documentation.* Paris: Gauthier-Villars, 1969. 286pp.
Lists and describes over 700 sources arranged under the following categories: bibliographies, general, demographic and health statistics, economic, cultural, political and legal statistics. An appendix lists bibliographical sources of national statistics.

**438 Harvey, Joan M.** *Sources of statistics.* 2nd edn. London: Clive Bingley; Hamden, Connecticut: Linnet Books, 1971. 126pp.
While emphasizing British sources it also gives international and American coverage on the following ranges of statistics: general; population; social; educational; labour; production; trade; finance; prices; transport; tourism.

**439** *Population index bibliography cumulated 1935–1968 by authors and geographical areas.* Boston, Massachusetts: Princeton University; G. K. Hall, 1971.
Contains no fewer than 173,000 entries. In Part 1 titles are arranged alphabetically by author. Official statistical publications are subdivided into census, vital statistics, yearbooks, others. In Part 2 titles are arranged by continent and country, then by major topics.

**440 Texas University, Population Research Center.** *International population census bibliography.* Austin, Texas: Bureau of Business Research, Texas University,1965–68. 7 vols.
The coverage is worldwide and the volumes are arranged according to area: 1. Latin America and the Caribbean; 2. Africa; 3. Oceania; 4. North America; 5. Asia; 6. Europe; 7. Supplement.

**441 United Nations Statistical Office.** *Supplement to the 'Statistical Yearbook' and the 'Monthly Bulletin of Statistics': methodology and definitions,* pp. 405–419. 2nd issue. New York: United Nations, 1972.
Its arrangement is alphabetical by countries, for which the following information is provided: central statistical agency; public or private statistical institutions; major series of each.

**442 U.S. Library of Congress Census Library Project.** *Statistical yearbooks: an annotated bibliography of the general statistical yearbooks of the major political subdivisions of the world.* Prepared by Phyllis G. Carter. Washington, D.C.: Card Division, Library of Congress, 1954. 93pp.
Arrangement by continent, then by country. Lists monthly and quarterly statistical bulletins.

**443 Wasserman, Paul** (managing ed.), **Joanne Paskar** (associate ed.). *Statistics sources: a subject guide to data on industrial, business, social, educational, financial, and other topics for the United States and internationally.* Detroit: Gale Research, 1974. 892pp.
Twenty-one thousand references arranged according to subject.

## Guides to statistics: national

### United Kingdom

**444 Central Statistical Office.** *List of principal statistical series publications.* London: HMSO, 1974. 52pp.
This list is regularly updated by the quarterly *Statistical News.*

**445** *Guides to official sources.* Interdepartmental Committee on Social and Economic Research. London: HMSO. Nos. 1–6.
These guides are titled as follows: No. 1. Labour statistics; No. 2. Census reports of Great Britain; No. 3. Local government statistics; No. 4. Agricultural and food statistics; No. 5. Social security statistics; No. 6. Census of production reports.

### Canada

**446 Dominion Bureau of Statistics Library.** Canada Yearbook Division. *Historical catalogue of Dominion Bureau of Statistics Publications, 1918–1960.* Bureau Fédéral de la Statistique, Division de l'Annuaire du Canada. *Catalogue rétrospectif des publications du Bureau Fédéral de la Statistique.* Ottawa: Queen's Printer.
This guide to the publications of the Dominion Bureau of Statistics is kept up to date by the issues of *Current Publications.*

### United States

**447** *American statistics index. 1974 annual and retrospective edition. A comprehensive guide and index to the statistical publications of the U.S. government.* Washington, D.C.: Congressional Information Service, 1974. 3 vols.
The major source for U.S. government statistical publications. Part 1 includes index by subject and name; index by category; guide to standard classifications; index by title; index by agency report number. Part 2 is arranged by department, bureau and series. It includes publications in print at 1 January 1974 and major publications issued since the early 1960s.

**448** *First annual supplement. Covering publications issued January 1–December 31, 1974.* 1975. 2 vols.
Part 1 consists of a 571-page index, Part 2 of 821 pages of abstracts.

**449 Andriot, John L.** *Guide to U.S. government statistics.* 4th edn. McLean, Virginia: Documents Index, 1973. 431pp.
A guide to over 1,700 statistical series arranged according to government department and bureau.

**450 U.S. Bureau of the Budget,** Office of Statistical Standards. *Statistical services of the United States.* Revised edn. Washington, D.C.: Government Printing Office. 156pp.

A survey of government services that has been reprinted in Andriot's guide [449] and is updated in *Statistical Reporter*, a monthly publication of the U.S. Office of Management and Budget.

**451 U.S. Bureau of the Census.** *Bureau of the Census Catalog of Publications, 1790–1972.* Washington, D.C.: Government Printing Office, 1974.

Part 1 (320pp.) lists a catalogue of U.S. census publications 1790–1945 and contains 3,664 entries. Part 2 (591pp.) lists Bureau of the Census publications 1946–1972.

**452 U.S. Bureau of the Census.** *Bureau of the Census Catalog, 1946–.* Washington, D.C.: Government Printing Office, 1947–.

Arrangement is by categories: general; agriculture; construction and housing; distribution and services; foreign trade; current trade with Puerto Rico and other U.S. possessions; geography; government; manufacturing and mineral industries; population; transportation.

## Statistical Cartography

### Bibliography

**453 Porter, Philip W.** *A bibliography of statistical cartography.* Minneapolis: University of Minnesota Bookstore, 1964. 64pp.

Contains 982 entries arranged alphabetically by author, topics covered being referenced in the margin, e.g. automation; graphs and diagrams; general methods and textbooks; population mapping; quantitative methods of spatial description; symbols. A useful reference aid for cartographers, geographers, librarians and anyone who is called upon to prepare statistical maps without professional assistance.

## Statistical applications in geography

### Bibliographies

**454 Greer-Wootten, Bryn.** *A bibliography of statistical applications in geography.* Washington, D.C.: Association of American Geographers, 1972. 91pp. (Commission on College Geography Technical Paper 9.)

Contains over 800 entries covering thirty-four types of statistical application in geography. For each section the bibliography is in alphabetical order.

**455 Anderson, Marc.** *A working bibliography of mathematical geography.* Ann Arbor: University of Michigan Department of Geography, 1963. 52pp. (Michigan Inter-University Community of Mathematical Geographers Discussion Paper 2.)

References are arranged alphabetically by author.

*General works*

**456 Dickinson, G. C.** *Statistical mapping and the presentation of statistics.* London: Edward Arnold, 1963. Reprinted 1967.
An undergraduate-level text on the presentation of statistical data in the form of maps and diagrams. Includes a list of census reports for Great Britain, 1801–1931. Clearly written, and illustrated in black and white.

**457 Gregory, Stanley.** *Statistical methods and the geographer.* 3rd edn. London: Longman, 1973. 271pp.
A basic text for geography students in higher education. Good bibliography on pp. 252–260.

**458 Norcliffe, G. B.** *Inferential statistics for geographers.* New York: John Wiley, 1977. 272pp.
Specifically written for first- and second-year students in higher education who are being introduced to statistics for the first time.

**459 Tomlinson, R. F.** *Geographical data handling.* Ottawa: International Geographical Union Commission on Geographical Data Sensing and Processing, 1972. 2 vols. 1,327pp.
An important work that includes chapters on data collection and processing; data manipulation; computer display applications; automated cartography; geographic information systems. A valuable reference work for geographers and cartographers.

# Cartographic Communication

*Bibliography*

**460 Board, C.** (ed.). *Bibliography of works on cartographic communication/ Bibliographie des oeuvres sur la communication cartographique.* International Cartographic Association/Association Cartographique Internationale, 1976. 147pp. (Available from the author, Department of Geography, London School of Economics, Houghton Street, Aldwych, London WC2.)
Contains 870 citations.

*Major works*

**461 Board, C.** (ed.). 'New insights in cartographic communication'. *Cartographica*, **21**(1) (1984). 138pp. (Cartographica Monograph No. 31.)
The Proceedings of the International Cartographic Association Seminar at the Royal Society, London.

**462 Guelke, Leonard** (ed.). 'The nature of cartographic communication'. *Cartographica* (1977). 147 pp. (Cartographica Monograph No. 19.)
Guelke brings together essays on the theme of maps as elements in a process of communication between mapmaker and map user. The work is divided into three parts: Part I provides a historical perspective; Part II contains five papers

showing how the major thrust of modern theoretical cartography has been in the application of ideas from communications theory to the analysis of maps; Part III consists of three papers raising objections to the idea of cartography as a communications science. The volume is important for professional cartographers, academic geographers and students. In adopting a communications theory of maps, however, cartographers have been influenced by other disciplines and there is a particular relationship with electronic communications theory.

### Selected papers

**463 Keates, J. S.** 'Cartographic communications: a critical review'. In *Understanding maps*, pp. 95–128. London and New York: Longman, 1982.
Keates discusses the numerous communication models involving information transmission and reception, then looks at map content, the cartographic language, map use, map design, cognition and perception, all in the context of cartographic communication. Generally of interest to professional mapmakers and academic geographers.

**464 Taylor, D. R. F.** (ed.). *Graphic communication and design.* Chichester: John Wiley, 1983. 332pp. (Vol. 2 in the series Progress in Contemporary Cartography.)
Assesses recent achievements in communication and design and examines new challenges such as designing for the computer.

Numerous papers have been written in recent years on various aspects of cartographic communication. Among them is the following, which will be mainly of interest to professional cartographers or geographers:

**465 Koeman, Cornelis.** 'The principle of communication in cartography'. *International Yearbook of Cartography*, pp. 169–176 (1971).

## Information Processing

## Map production

### Major works

**466 Cuenin, V.** *Cartographie générale:* Tome 2, *Methodes et techniques de production.* Paris: Editions Eyrolles, 1973.
This volume is divided into two parts, the first being concerned with map reproduction (mainly offset lithography); it includes a useful chapter on process photography. The second part covers compilation and preparation including selection and generalization, drafting, colour separations and automation. The text is in French. Useful for students of cartography and geography.

**467 Keates, J. S.** *Cartographic design and production.* London: Longman, 1973. 240pp.

The standard work on map production. Suitable as a standard text for HNC/ HND or degree courses in cartography, surveying or topographic science. Useful also as a reference work for students of related disciplines such as photogrammetry, geography and the earth sciences. Has sections on process photography; non-photographic systems such as dyeline; multiple reproduction processes; proofing processes; automation of image formation; production planning. There is a good bibliography on pp. 231–233.

**468 Loxton, John.** *Practical map production.* New York, Chichester, Santa Barbara, Brisbane and Toronto: John Wiley, 1980. 137pp.
Designed as an aid to students and teachers and as a reference manual for land surveyors. Includes chapters on planning; drafting; in the cartographic drawing office; map revision; computers and cartography.

### Other works and selected papers

**469 Dennett, J. R. B., L. H. E. Hobbs** and **B. F. White.** 'Cartographic production control'. *The Cartographic Journal*, **4**(2): 93–103 (1967).
Especially aimed at professional cartographers.

**470 Hulbert, F.** and **S. E. Tischler.** 'Contrôle de la production en cartographie'. *International Yearbook of Cartography*, **5**: 113–126 (1965).

**471 Leatherdale, J. D.** and **K. M. Keir.** 'Digital methods of map production'. *Photogrammetry Record*, **9**(54): 757–778 (1979).
Primarily for photogrammetrists and cartographers.

**472 Render, J.** *Map preparation: some guidance on fundamentals.* 2nd edn. (Available from the author, Department of Geography, Portsmouth Polytechnic, Lion Terrace, Portsmouth PO1 3HE, England.) 1978. 57pp.
A useful introduction for student cartographers and young entrants to the profession.

**473 Tanaka, K.** and **Y. Inamura.** 'Geological maps published by the Geological Survey of Japan—from drafting and printing up to publication'. *International Yearbook of Cartography*, **20**: 170–179 (1980).
The authors discuss the complete process of map production with specific reference to geological maps but their paper is also of interest to geographers, cartographers and earth scientists generally.

## Map projections

Many specialized works are available on the construction of map projections. Among the most popular are the following:

**474 Kellaway, G. P.** *Map projections.* London: Methuen, 1970. 127pp.
An introduction to the subject that is suitable for sixth-form students or first-year university courses in cartography.

**475 McDonnell, Porter.** *Introduction to map projections.* New York and Basle: Marcel Dekker, 1979. 174pp.
An introductory work which is suitable for sixth forms and first-year courses in higher education. Clearly written and containing a large number of exercises, some of which are worked examples in the text, while others set problems for the reader.

**476 Maling, D. H.** *Coordinate systems and map projections.* London: George Philip & Son, 1973. 255pp.
A more advanced work than [474] suitable for specialized courses in cartography and university courses in geography.

**477 Snyder, J. P.** *Map projections used by the Geological Survey.* 2nd edn. U.S. Geological Survey Bulletin No. 1532. Washington, D.C.: Government Printing Office, 1983. 313pp.

**478 Steers, J. A.** *An introduction to the study of map projections.* 14th edn. London: University of London Press, 1965. 292pp.
Probably the most widely used work on projections. Suitable for first-year university courses in geography and for specialized cartography courses.

Map projections are also dealt with in the standard general works on cartography and in numerous papers in cartographic journals; for example, Jean M. Ray tables the properties of map projections and provides a list of references in Special Libraries Association, Geography and Map Division, *Bulletin,* **105**: 6–11 (September 1976).

## Thematic maps: preparation and production

Virtually all the standard texts on modern mapmaking devote considerable space to thematic maps and numerous articles in the cartographic and geographic journals are devoted to various aspects of the subject. Among the most popular works are:

**479 Cuff, David J.** and **Mark T. Mattson.** *Thematic maps: their design and production.* New York and London: Methuen, 1982. 169pp.
A very practical book by authors who combine academic and professional expertise. Designed as an introduction to the principles and practices of thematic mapping for first-year courses in geography and cartography. Very suitable for the beginner in professional cartography. Up to date in its discussion of equipment and practices. Has an excellent bibliography.

**480 Dickinson, G. C.** *Statistical mapping and the presentation of statistics.* 2nd edn. London: Edward Arnold, 1973.
Written by a university geographer. Provides a clear introduction to graphical methods of statistical presentation.

**481  Hodgkiss, A. G.** *Maps for books and theses.* Newton Abbot: David & Charles;
New York: Pica Press, 1970. 267pp.
Designed as a practical guide to the preparation of thematic maps for geography
and cartography students or for anyone who has to prepare maps and diagrams
for illustrative purposes.

**482  Lawrence, G. R. P.** *Cartographic methods.* 2nd edn. London: Methuen, 1979.
153pp.
Provides a general survey of those aspects of cartography which are most
important to geographers. Suitable for first-year courses in departments of
geography.

**483  Lewis, Peter.** *Maps and statistics.* New York: John Wiley, 1977. 318pp.
Proceeds from an introduction of measurement and probability to the analysis of
point symbol and line symbol maps. Each section has its own exercises with
answers provided in an appendix. Good introduction to statistics for geographers
and others.

**484  Monkhouse, F. J.** and **H. R. Wilkinson.** *Maps and diagrams; their compilation and construction.* 3rd edn. London: Methuen, 1973.
The first major post-war textbook on statistical mapmaking. Deals with a great
variety of data, types of map and methods of presentation. Remains a valuable
source of reference for cartographers, geographers and other scientists called
upon to prepare maps for papers, books or dissertations.

**485  Robinson, Arthur H., Randall Sale** and **Joel Morrison.** *Elements of
cartography.* 4th edn. New York, Santa Barbara, Chichester, Brisbane and
Toronto: John Wiley, 1978. 448pp.
A comprehensive guide covering all aspects of the subject. Looks at the historical
background; mathematical geography; data acquisition, presentation and compilation; remote sensing; generalization; symbolization; computer-assisted cartography; colour; typography and lettering; graphic design. The standard work
for students of cartography and a useful reference aid or textbook for geographers and any earth scientists who are required to prepare their own maps.

# Map design

Map design is a topic that is dealt with in a great many papers in the
cartographic journals and in a small number of books. It is also studied to
varying degrees in the general textbooks on cartography.

## Bibliography

**486  Brod, Raymond M.** 'A design bibliography for cartographers'. Special
Libraries Association, Geography and Map Division, *Bulletin*, **104**: 27–33
(June 1976).
Cites ninety-seven references to textbooks and readily available articles that are

likely to assist professional cartographers, students, geographers and workers in any discipline who prepare maps for papers, books or dissertations.

## Major works

**487 Castner, H. W.** and **C. McGrath** (eds.). *Map design and the map user.* Toronto: York University, 1971. 84pp. (Cartographica Monograph No. 2.)
An important collection of essays on aspects of map design. Essential reading for the professional cartographer but also of value to cartography and geography students and anyone required to prepare maps for dissertation or other purposes.

**488 Keates, J. S.** *Cartographic design and production.* London: Longman, 1973. 240pp.
The standard work for teachers and students of cartography, surveying or topographic science. A valuable reference aid.

**489 Kingsbury, R. C.** *Creative cartography: an introduction to effective thematic map design.* Bloomington, Indiana: Indiana University, 1969. 44pp.
A brief introduction to map design for students of geography and related disciplines.

**490 Robinson, Arthur H.** *The look of maps: an examination of cartographic design.* Madison, Wisconsin: University of Wisconsin Press, 1952. 105pp.
A specialized treatise on map design that would be valuable mainly to professional cartographers and possibly to graphic designers who occasionally prepare maps.

## Papers

**491 Board, C.** and **R. M. Taylor.** 'Perception and maps: human factors in map design and interpretation'. *Transactions* of the Institute of British Geographers, New Series, 2(1) (1977).

**492 Flanders, D. W.** 'The design and preparation of maps for 35mm projectors'. *The Cartographic Journal*, **13**(1): 89–93 (June 1976).
A useful guide for academic geographers, educational cartographers, school teachers and graphic artists.

**493 Irwin, D.** 'Black and white maps from color'. *The Canadian Cartographer*, **8**(2): 137–142 (December 1971).
Gives advice on the preparation of derived black and white maps for cartographers working in educational establishments, geographers, and earth and social scientists.

**494 Odell, C.** 'Design requirements for classroom maps'. *Papers* from the 28th Annual Meeting of the American Congress on Surveying and Mapping, Washington, D.C., 1968, pp. 176–181.
Lecturers and school teachers will find useful advice.

**495 Raisz, Erwin.** 'Draw your own blackboard maps'. *Journal of Geography*, **XLI** (1942).

**496 Robinson, Arthur H.** 'Psychological aspects of color in cartography'. *International Yearbook of Cartography*, pp. 50–61(1967).
A specialized paper for professional mapmakers.

**497 Sandford, H.** 'Map design for children'. *SUC Bulletin*, **14**(1): 39–48 (1980).
Provides helpful advice for teachers, geographers and compilers of school atlases.

**498 Sen Gupta, A.** 'Concepts of legends for thematic maps'. *Geographical Review of India*, **35**: 43–51 (March 1973).

**499 Skop, J.** 'The effect of user requirements on map design'. *Surveying and Mapping*, **18**: 315–318 (1958).

**500 Wood, Michael.** 'Visual perception and map design'. *The Cartographic Journal*, **5**: 54–64 (1968).

## Symbols, tones, patterns

The topic of symbols, tones and patterns is dealt with in the general works on map preparation but there are also many papers and articles covering more specialized aspects. Some examples are given below.

**501 Alexandra, J. W.** and **G. A. Zahorchak.** 'Population density maps of the United States: techniques and patterns'. *Geographical Review*, **XXXIII**: 457–466 (1943).
Useful to all who prepare maps, for the principles applied to population density mapping apply to a host of other data.

**502 Crawford, P. V.** 'Perception of grey-tone symbols'. *Annals* of the Association of American Geographers, **61**: 721–735 (1971).

**503 Crawford, P. V.** 'The perception of graduated squares as cartographic symbols'. *The Cartographic Journal*, **10**: 85–94 (December 1973).
Squares are less commonly used than circles as graduated symbols in thematic mapping. Crawford examines their effectiveness.

**504 Cuff, D. J.** 'Shading on choropleth maps'. *Proceedings* of the Association of American Geographers, **5**: 50–54 (1973).
The success of any choropleth map depends to a great extent on the choice of an effective shading system. Cuff's paper is therefore important to all thematic mapmakers, whether amateur or professional.

**505 Dobson, M.** 'Symbol–subject matter relationships in thematic cartography'. *The Canadian Cartographer*, **12**: 52–67 (June 1975).

**506 Dreyfuss, H.** *Symbol sourcebook*. London: McGraw-Hill, 1972. 292pp.
Pages 94–97 of this general work on symbols are devoted to geography and
geology. A useful reference for geographers, geologists and earth scientists who
prepare their own maps.

**507 Head, G.** 'Land–water differentiation in black and white cartography'. *The
Canadian Cartographer*, **9**(1): 25–38 (June 1972).
The clear differentiation between land and water is an important factor in
successful map design, particularly in maps used for illustrative purposes in books
and journals. This is a useful paper for geographers, students of cartography, etc.

**508 Jenks, G. F.** and **D. S. Knos.** 'The use of shaded patterns in graded series'.
*Annals* of the Association of American Geographers, **51**: 316–334 (1973).
One of most frequently encountered problems in designing thematic maps is the
construction of a graded series of patterns. Jenks and Knos provide a useful
examination of the problem.

**509 Keates, J. S.** 'Symbols and meaning in topographic maps'. *International
Yearbook of Cartography*, **12**: 168–181 (1972).
Keates looks at symbolization in topographic rather than thematic maps in a
paper that is useful to both mapmaker and map user.

**510 Kilcoyne, J.** 'Pictographic symbols in cartography'. *Proceedings* of the
Association of American Geographers, **6**: 87–90 (1974).

**511 Lehmann, E.** 'Symbol systems in thematic cartography'. *International
Yearbook of Cartography*, pp. 28–31 (1972).

**512 Meihoefer, H.** 'The visual perception of the circle in thematic maps;
experimental results'. *The Canadian Cartographer*, **10**: 63–84 (June 1973).
A revealing look at the results of using graduated circles for all who make or use
statistical maps, particularly geographers and social scientists.

**513 Sen Gupta, A.** 'Symbolisation of information in thematic mapping'.
*Geographical Review of India*, **36**: 145–154 (June 1974).
Mainly useful for geographers.

**514 Williams, R.** 'Map symbols: equal appearing intervals for printed screens'.
*Annals* of the Association of American Geographers, **48**: 132–139 (June 1958).
A technical article mainly of use to professional cartographers.

**515 Williams, R.** 'Statistical symbols for maps: their design and relative
values'. Washington, D.C.: Office of Naval Research, Contract NR008–006
Nonr 609 (03).

**516 Wilson, E.** 'The design and selection of graded shadings for black and
white maps'. *SUC Bulletin*, **1**: 11–16 (1967).
Written by a professional cartographer for younger colleagues.

# Simplification and generalization

Simplification and generalization is a particularly important concept in all kinds of mapwork. The following essay includes a very useful bibliography.

**517 Steward, H. J.** *Cartographic generalisation: some concepts and explanation.* Toronto: York University, 1974. 78pp. (Cartographica Monograph No. 10.)
This essay reviews general thinking about cartographic generalization and sets the scene for a quantitative examination. Steward provides a useful bibliography of some 270 references on pp. 50–77. Essential reading for professional and amateur mapmakers.

## Selected papers

**518 Das Gupta, S. P.** 'Some measures of generalization on thematic maps'. *Geographical Review of India,* **26**(2): (1964).
A helpful guide for geographers, statistical cartographers and student mapmakers.

**519 Das Gupta, S. P.** 'Cartographic generalization as applied to thematic mapmaking'. Paper given at the IGU/ICA meeting in New Delhi in 1968. 11pp.

**520 Delaney, G. F.** 'Name generalization'. *Canadian Cartography,* **1**(1): 22–24 (1962).

**521 Fahey, Lawrence.** *Generalization as applied to cartography.* Unpublished M.A. thesis, Ohio State University, 1954. 180pp.

**522 Floyd, A. M.** 'Generalisation problems in derived mapping'. *Canadian Cartography,* **1**(1): 4–7 (1962).
A discussion of generalization as concerned with small-scale thematic mapping. Helpful for anyone called upon to prepare maps without specialized assistance.

**523 Jenks, George C.** 'Generalization in statistical mapping', *Annals* of the Association of American Geographers, **53**(1): 15–26 (1963).

**524 Knorr, H.** 'Generalisation, revision and automation'. *Nachrichten aus dem Karten- und Vermessungswesen,* Reihe V, No. 4: 7–21 (1963).

**525 Koeman, Cornelis** and **F. L. T. van der Eiden.** 'The application of computation and automatic drawing instruments to structural generalisation'. *Cartographic Journal,* **7**(1): 47–48 (June 1970).
A technical article mainly for professional cartographers.

**526 Long, D. E.** 'Generalisation of culture in derived mapping'. *Canadian Cartography,* **1**(1): 12–17 (1962).

**527 Lundquist, G.** 'Generalization—a preliminary survey of an important subject'. *Canadian Surveyor*, **14**(10): 466–470 (1959).
Looks at generalization of topographic maps for surveyors and professional cartographers.

**528 Lundquist, G.** 'Generalization of communication networks'. *Nachrichten aus dem Karten- und Vermessungswesen*, Reihe V, No. 5: 35–42 (1963).

**529 Mallet, M. J.** 'Problèmes de généralisation (données qualitatives et quantitatives) et mise à jour en cartographie thématique'. *Nachrichten aus dem Karten- und Vermessungswesen*, Reihe V, No. 4: 107–112 (1963).
Looks at generalization of topographic as well as thematic maps.

**530 Miller, O. M.** and **Robert J. Voskuil.** 'Thematic map generalization'. *Geographical Review*, **54**(1): 13–19 (January 1964).

**531 Morrison, Joel L.** 'A theoretical framework for cartographic generalization with emphasis on the process of symbolization'. *International Yearbook of Cartography*, **XIV**: 115–127 (1974).

**532 Rhind, D. W.** 'Generalisation and realism within automated cartographic systems'. *Canadian Cartographer*, **10**(1): 51–62 (June 1973).
Aimed primarily at professional cartographers but also of interest to geographers.

**533 Robinson, Arthur H., Randall Sale** and **Joel Morrison.** 'Cartographic generalization'. In *Elements of cartography* [485], pp. 149–180.
Chapter 8 of this standard text is a lengthy discussion of generalization and is arranged under the following headings: Generalization; The elements of generalization; The operation of the controls of generalization; Common simplification and classification; Data manipulations. A very useful essay that should be read by all who prepare their own maps as well as by professional cartographers and geographers.

## Calligraphy; typography

Calligraphy is dealt with to some degree in the standard works but the authors are not always specialists and it can sometimes be more helpful to turn to one of the many works that deal exclusively with the subject—though not (except in the case of Heather Child's books) with its applications to cartography.

**534 Child, Heather.** *Calligraphy today*. London: Studio Vista, 1976. 112pp.
This book reviews the work produced by outstanding British, European and American calligraphers and also gives a brief history of the craft during the twentieth century. The illustrations are carefully chosen to demonstrate the wide variety of purposes to which calligraphy is put, including decorative maps and maps drawn as book illustrations.

**535 Child, Heather.** *Decorative maps.* London and New York: Studio Publications, 1956. 96pp.
This, No. 61 in a series of 'how to do it' publications, is written by a distinguished calligrapher and shows the evolution of the map from an idea to the finished design. Hand lettering is dealt with on pp. 51–54.

**536 Gates, David.** *Lettering for reproduction.* New York: Watson-Guptill Publications, 4th printing, 1976. London: Pitman, 1976. 192pp.
Designed primarily for students of graphic design, it is a valuable work for anyone who is required to prepare maps and diagrams for illustrative purposes. Deals with tools, materials and equipment as well as providing instruction in the different styles and methods of lettering.

Among the numerous articles dealing with the lettering of maps are the following:

**537 Dawson, W.** 'The lettering of maps'. *Cartography,* **1**: 85–88 (1955).

**538 Riddeford, C.** 'On the lettering of maps'. *The Professional Geographer,* **4**: 7–10 (September 1962).

**539 Robinson, Arthur H.** 'The size of lettering for maps and charts'. *Surveying and Mapping,* **10**: 37–44 (1950).
A useful article for those preparing maps for illustrative use.

**540 Withycombe, J. G.** 'Lettering on maps'. *The Geographical Journal,* **73**(5): 429–446 (1929).

## Typography

Among the many volumes dealing with typography the following will be particularly useful:

**541 Perfect, Christopher** and **Gordon Rookledge.** *Rookledge's International type-finder: the essential handbook of typeface recognition and selection.* London: Sarema Press, 1983. 270pp.
A valuable working tool for designers, graphic artists, and a reference for cartographers, publishers and editors that should prove indispensable.

**542 Swann, Cal.** *Techniques of typography.* London: Lund Humphries, 1969. 96pp.
A useful textbook and reference for printing apprentices, design students, graphic artists, professional cartographers.

## Positioning names on maps

**543 Imhof, Eduard.** 'Positioning names on maps'. *The American Cartographer,* **2**(2): 128–144 (1975).

A well-illustrated article in which Professor Imhof states that legibility and clarity of maps depend to a large extent on the judicious positioning of names with one optimum position in which each individual name should be placed. An important reference for students of cartography, geography and all disciplines involving mapwork.

# Computer-assisted cartography

## Glossary

**544 Edson, D. T.** (ed.). *Glossary of terms in computer assisted cartography/Glossaire des termes en cartographie assistée par ordinateur.* 2nd edn. Falls Church, Virginia: American Congress on Surveying and Mapping, for the International Cartographic Association/Association Cartographique Internationale, 1980. 157pp.
English/French glossary. A necessary reference work for all using computer-assisted cartography.

## Report

**545 Starr, L.** (compiler). *Computer-assisted cartography: research and development report.* International Cartographic Association/Association Cartographique Internationale, 1984. 124pp. (Available from R. Muller, P.O. Box 9016, NL-1006 AA Amsterdam, The Netherlands.)

## Introduction to computer graphics

**546 Angell, Ian O.** *A practical introduction to computer graphics.* London: Macmillan, 1981.
An introduction that is mainly concerned with the display of fairly simple geometrical forms. A number of the techniques described have some relevance to cartography.

## Selected works

**547 Fraser Taylor, D. R.** (ed.). *Progress in Contemporary Cartography;* Vol. 1, *the computer in contemporary cartography.* Chichester: John Wiley, 1980. 252pp.
Eminent workers in the field have produced a book in which attention is focused on various aspects of computer-assisted mapmaking. These include: the impact of computer technology on cartography; the nature of computer-assisted cartography; latest developments in techniques and equipment; special applications; the impact of computers on Swedish mapping; geological maps; census maps; GIMMS. There are references at the end of each chapter. A useful review of the state of computer mapping at the close of the 1970s.

**548 Monmonier, Mark S.** *Computer-assisted cartography: principles and prospects.* Englewood Cliffs, New Jersey: Prentice-Hall International, 1982. 214pp.
The text introduces the reader to computer procedures, and includes sections on map projections and cartographic data structures, computer-assisted map design and the planning of colour maps. Designed primarily for geographers and cartographers. It includes a helpful list of selected literature.

*Other works: selected papers, etc.*

**549 British Cartographic Society** Special Publication No. 2: *Computers in cartography*. 1983. Available from A. G. Williams, 39 North Street, Winterbourne Stickland, Blandford Forum, Dorset DT11 0NJ, England.

**550 Carter, J. R.** 'Computer mappers and map librarians—can they help each other?' Special Libraries Association, Geography and Map Division, *Bulletin*, No. 113: 49–64 (September 1978).
Designed to help map librarians in their role as information specialists. Includes a list of references and an aid to vocabulary building in computer mapping.

**551 Cartwright, R. S.** *Reference manual for Program CALFORM*. Laboratory for Computer Graphics and Spatial Analysis, Graduate School of Design, Harvard University, 1972.

**552 Castle, Dorothy.** 'The computer and the cartographer'. *SUC Bulletin*, **6**(2): 25–27 (March 1972).
Basic introduction for undergraduates and others who are required to prepare thematic maps.

**553 Davis, John C.** and **M. J. McCullagh.** *Display and analysis of spatial data*. Chichester, New York, Toronto, Brisbane and Santa Barbara: John Wiley, 1975.
This volume is the collective result of the 1973 NATO Advanced Study Institute Conference and provides an excellent overview of the field.

**554 Dawson, J. A.** and **D. J. Unwin.** *Computing for geographers*. Newton Abbot: David & Charles, 1976. 362pp.
A practical work aimed primarily at geographers but of use to other earth scientists and mapmakers.

**555 Douglas, D. H.** 'Mapmaking with the electronic digital computer'. In H. M. French and J-B. Racine (eds.) *Quantitative and qualitative geography*, pp. 97–114. Ottawa: University of Ottawa, Department of Geography Occasional Paper, 1971.

**556 Fisher, H. T.** *et al. Reference manual for Synographic Computer Mapping (SYMAP) Version V*. Cambridge, Massachusetts: Laboratory for Computer Graphics and Spatial Analysis, Graduate School of Design, Harvard University, 1968.
The basic manual for one of the most popular thematic mapping programs.

**557 Gaits, G. M.** 'Thematic mapping by computer'. *The Cartographic Journal*, **6**(1): 50–68 (1969).
A useful introduction for students of geography and cartography as well as statistical mapmakers generally.

**558 Greggor, K. N.** 'Computer-assisted map compilation'. *Cartography*, **9**(1): 24–34 (1975).

**559 Harvard Computer Mapping Collection**: The Standard Library, vol. 1. *Managements' use of maps: commercial and political applications*. Chestnut Hill, Massachusetts: Harvard Library for Computer Graphics, 1980. 72pp.
Eight case studies ranging from the use of computer mapping by Congress and the White House to commercial firms.

**560 Hsu, M** and **P. W. Porter.** 'Computer mapping and geographical cartography'. *Annals* of the Association of American Geographers, **61**: 797–799 (1971).

**561 Jeffery, M., H. O'Hare** and **C. Board.** 'Choropleth maps on the microfilm plotter: an attempt to improve the graphic quality of automated maps'. *International Yearbook of Cartography*, **15**: 39–46 (1975).

**562 Leatherdale, J. D.** and **K. M. Keir.** 'Digital methods of map production'. *Photogrammetry Record*, **9**(54): 757–758 (1979).
A short introduction for cartographers and photogrammetrists.

**563 LeBlanc, Aubrey L.** (ed.). *Computer cartography in Canada*. Toronto: York University, 1973. 103pp. (Cartographica Monograph No. 9.)
A collection of papers covering differing Canadian applications of computer cartography; an approach to automatic cartography for topographic mapping; automated cartography in Canadian Federal mapping; automating marine chart production; a semi-automated production system for engineering plans; the relation of automatic mapping to urban information systems; conformal representation of the Prairie Provinces; present trends and future challenges.

**564 Liebenberg, E.** 'SYMAP: its uses and abuses'. *The Cartographic Journal*, **13**: 26–35 (1976).
Some applications of one of the most popular programs.

**565 Margerison, T.** *Computers and the renaissance of cartography*. London: Experimental Cartographic Unit, Royal College of Art, 1976. 20pp.
Summarizes developments over the ten years of the unit's existence. Discusses recent applications and suggests future guidelines.

**566 McCormack, D. D.** 'Computers and cartography'. *New Zealand Cartographic Journal*, **11**(1): 5–14 (1981).

**567 Monmonier, M. S.** 'The scope of computer mapping'. Special Libraries Association, Geography and Map Division, *Bulletin*, No. 18: 2–14 (September 1970).
Primarily intended to introduce librarians and map curators to developments in and applications of computer-assisted cartography.

**568 Reilly, W. J.** 'Computer-assisted contour mapping'. *New Zealand Cartographic Journal*, **11**(1): (1981).
For professional cartographers.

**569 Rentmeester, L. F.** 'United States Department of Defense Cartographic Data Handling System'. *The Cartographer*, **3**(2): 127–137 (December 1966).
Discusses automation of the data handling system—a system that is compatible with current planning for an improved national network of information systems in science and technology.

**570 Scripter, M. W.** 'Choropleth maps on small digital computers'. *Proceedings of the Association of American Geographers*, **1**: 133–136 (1969).
Helpful for anyone preparing maps showing densities per unit of area and particularly for geographers, thematic mapmakers, sociologists, economists, etc.

**571 Stutz, F. P.** 'World map projections on the computer'. *The Canadian Cartographer*, **11**: 56–68 (1974).
Mainly of interest to professional mapmakers, cartography students and geographers.

**572 Wastesson, Olof, Bengt Rystedt** and **D. R. F. Taylor.** *Computer cartography in Sweden*. Toronto: York University, 1977. (Cartographica Monograph No. 20.)
Eleven papers from a Swedish Seminar on Geographic Information Systems held at Gävle in 1977 and containing much to interest planners, local government officers, geographers and cartographers. Topics covered include the Swedish Land Data Bank; the Swedish Road Data Bank; the Geocoded Data System of Malmö Municipality; information systems at the National Land Survey of Sweden; and geographic information systems in Sweden.

**573 Yoeli, P.** 'Cartographic drawing with computers'. *Computer Applications*, vol. 8, 1982. 137pp. (Available from the Department of Geography, The University, Nottingham NG7 2RD, England.

## Atlases

**574 Alexander, G. L.** *Guide to atlases: world, regional, national, thematic. An international listing of atlases published since 1950*. Metuchen, New Jersey: Scarecrow Press, 1971. 671pp.

**575 Alexander, G. L.** *Guide to atlases supplement: world, regional, national, thematic. An international listing of atlases published 1971 through 1975 with comprehensive indexes*. Metuchen, New Jersey: Scarecrow Press, 1977. 362pp.
These two important volumes by Alexander are a valuable reference tool for librarians, map curators and map users in any discipline. The 8,550 entries list the name of the publisher, title of work, edition, author, place and date of publication, number of pages, and size.

**576 Lock, C. B. Muriel.** *Modern maps and atlases: an outline guide to twentieth century production*. London: Clive Bingley; Hamden, Connecticut: Archon Books, 1969. 619pp.
Dr. Lock provides information in narrative form on both topographic and

thematic maps and atlases. Information concerning topographic maps is treated regionally and that about thematic maps under subject matter. Has a full alphabetical index. An essential reference work for all librarians and map curators.

**577 Stams, W.** *National and regional atlases: a bibliographic survey.* International Cartographic Association/Association Cartographique Internationale, 1984. 250pp. (Available from R. Muller, P.O. Box 9016, NL-1006 AA Amsterdam, The Netherlands.)

**578 Winch, Kenneth L.** *International maps and atlases in print.* 2nd edn. London and New York: R. R. Bowker, 1976. 866pp.
Compiled by Kenneth Winch, who had previously compiled *Stanford's Reference guide*, this massive work includes material not easily accessible elsewhere. Entries are arranged by world, region and country. Within each country the arrangement is by theme. Winch also includes many map-index diagrams that are essential reference material. With its eight thousand entries this work is a *vade mecum* for map curators and should provide the information necessary to answer virtually any question about the availability of maps in print in 1976.

## Specialized atlases

**579 Stommel, Henry** and **Michele Fieux.** *Oceanographic atlases: a guide to their geographic coverage and contents.* Woods Hole, Massachusetts: Woods Hole Press, 1978. 97pp.
A useful reference work for oceanographers, hydrographers and geographers.

## Use of atlases

**580 Fremlin, G.** and **L. M. Sebert.** *National atlases.* Toronto: York University, 1972. 81pp. (Cartographica Monograph No. 4).
Discusses national atlases' history, analysis and ways towards improvement and standardization. A translation of *Atlas nationaux: analyse, voies de perfectionnement et d'unification* (prepared under the direction of Professor K. A. Salichtchev). The translation was specially made to facilitate the understanding of the principles of national atlases for the officers of the Surveys and Mapping Branch, Canada. Of value to editors, cartographers, historians and geographers.

**581** *The purpose and use of national and regional atlases.* Toronto: York University, 1979. 100pp. (Cartographica Monograph No. 23.)
A collection of papers presented at an international seminar held in Ottawa, March 1979. Examines the nature and value of national and regional atlases; the purpose and use of national atlases; the national atlases of the United States and Canada; the Junior Atlas of Alberta; computer techniques; and urban atlases. An important reference work for editors, cartographers and geographers.

# Map Use

## Major works

**582 Castner, H. W.** and **G. McGrath** (eds.). *Map design and the map user.*
Toronto: York University, 1971. 84pp. (Cartographica Monograph No. 2.)
This collection of essays includes such topics as 'The topographic map and the
map user', 'Mapping expedients to meet military requirements', 'Problems in
satisfying the needs of the Canadian 1:50,000 users', 'Published map use in a
consulting engineering office', 'Teachers as map users', 'A recreation map of the
central Cairngorms'. Prepared mainly for professional cartographers, the
volume is of interest to all map users.

**583 Hodgkiss, A. G.** *Understanding maps: A systematic history of their use and
development.* Folkestone: Wm. Dawson & Sons, 1981. 209pp.
The central theme is the art of using maps as a means of visual communication.
Well illustrated with examples of a wide range of map types. Covers modern as
well as historic maps. Aimed at all map users, professional and amateur.

**584 Muehrcke, Phillip C.** *Map use: reading, analysis and interpretation.* Madison,
Wisconsin: JP Publications, 1978. 469pp.
Muehrcke's book is highly readable by all map users, being written in a lively
style with pertinent illustrations. It has five sections: map reading; map analysis;
map interpretation; orientation; appendices (sources of maps; remote sensing;
map projections; enlarging and reducing; map care; tables). Provides useful and
stimulating reading for students of geography and the earth sciences as well as
for cartographers and the general reader.

**585 Barbour, Michael.** 'The use of maps: the topographic and thematic
traditions contrasted'. *The Cartographic Journal*, **20**(2): 76–86 (December 1983).
A study of maps used to illustrated printed books on geographical topics from
1500 to modern times.

**586 Dale, P. F.** 'Childrens' reactions to maps and aerial photographs'. *Area*,
**3**(3): 170–177 (1971).
Of particular interest to teachers, cartographers, editors, publishers and
designers.

**587 Drewitt, B.** 'The changing profile of the map user in Great Britain'. *The
Cartographic Journal*, **7**(1): 39 (June 1970).

**588 Kirby, R. P.** 'A survey of map user practices and requirements'. *The
Cartographic Journal*, **7**(1): 39 (1970).

**589 McCullagh, M. J.** and **R. J. Sampson.** 'User desires and graphics
capability in the academic environment'. *The Cartographic Journal*, **9**(2):
109–122 (1972).

An attempt to determine the average geographer's needs in relation to his research requirements and computing knowledge in automated map production. Designed for academic geographers, particularly those seeking to use computer-assisted cartography.

**590 Monmonier, M. S.** *Maps, distortion and meaning.* Washington, D.C.: Association of American Geographers, 1977. 51pp.

The map user is given some understanding of the theoretical basis of cartographic methods, symbol systems and design. The book stresses the importance of both map reader and map author playing an active role in cartographic communication. The emphasis is on small-scale thematic maps.

**591 North, Gary W.** 'Maps: who uses them?'. Special Libraries Association, Geography and Map Division, *Bulletin*, **130**: 2–18 (December 1982).

North's general emphasis is on American map usage. He discusses maps in today's information society; how many maps exist; what types are available; who uses them. Tables show production by different American establishments, categories of map user, etc. An interesting and useful reference for map curators, librarians, cartographers, geographers.

**592 Ordnance Survey** Professional Paper, New Series No. 28. *The map market in Great Britain.*

A survey carried out by the British Market Research Bureau and the Office of Population, Censuses and Surveys on the use of maps by householders in Britain.

## Map Reading

There is a great variety of literature at all levels on the subject of map reading. Among the better items are two works designed for military use:

**593** *Manual of map reading.* London: Ministry of Defence; HMSO, 1973.

Designed for use by army instructors but valuable for anyone involved in teaching map reading.

**594** *Map reading.* Washington, D.C.: Department of the Army, 1969.

Designed for use in the U.S. Army but a useful textbook for general use. Has four folded maps in a pocket.

Books on map reading suitable for schoolroom use at different levels include:

**595 Jennings, J. N.** and **S. O. Odnah.** *Certificate map interpretation.* Cambridge: Cambridge University Press, 1976.

Suitable for fifth and sixth forms.

**596 Matkin, R. B.** *Your book of maps and map reading.* London: Faber & Faber, 1970. 96pp.

Suitable for age levels ten to fourteen. A well-illustrated introduction to the topic.

**597 Meux, A. H.** *Reading topographical maps.* London: University of London Press, 1960.
A general introduction suitable for fifth and sixth forms.

Books that are suitable for sixth-form use and for undergraduate teaching include:

**598 Muehrcke, Phillip C.** *Map use: reading, analysis and interpretation.* Madison, Wisconsin: JP Publications, 1978. 469pp.
Part I of Muehrcke's book consists of seven chapters under the general heading of 'Map reading'. Topics covered are: reality transformed; mapping techniques; horizontal position; vertical position; temporal position; direction; distance. An excellent introduction to map appreciation for students and the general reader.

## Map Interpretation

**599 Dury, G. H.** *Map interpretation.* London: Pitman, 1960. 209pp.
An advanced textbook which assumes that the reader has some basic knowledge of the cartographic language. The emphasis is on physical geography.

## Map Care

### Bibliographies

**600 Piquette, Constance M.** 'Map librarianship: a selected and annotated bibliography'. Santa Cruz, California: Western Association of Map Libraries, *Information Bulletin*, **15**(1): 67–73 (1983).

**601 Schorr, Alan E.** 'Map librarianship, map libraries and maps: a bibliography, 1921–1973'. Special Libraries Association, Geography and Map Division, *Bulletin*, **95**: 2–35 (1974). Supplement in *Bulletin* **107**: 2–18 (1977).
Contains 813 references extracted from 'Library literature' 1921–1976.

**602 Woods, Bill M.** *Map librarianship: a selected bibliography.* New York: New Jersey Library Association, 1971. 20pp.

### Major works

**603 Drazniowsky, Roman** (compiler). *Map librarianship: readings.* Metuchen, New Jersey: Scarecrow Press, 1975. 548pp.
A collection of forty-eight papers originally published in various journals between 1950 and 1972 on all aspects of the care of maps. Provides useful background but overtaken by developments, particularly in computer-aided techniques.

**604 Ehrenberg, Ralph E.** *Archives and manuscripts: maps and architectural drawings.*
Chicago: Society of American Archivists, 1982. 64pp.
A useful introduction for archivists who lack specialist knowledge of maps and
drawings.

**605 Farrell, Barbara** and **Desbarats, Aileen.** *Guide for a small map collection.*
2nd edn. Ottawa: Association of Canadian Map Libraries, 1984. 88pp.
Gives clear and concise guidelines aimed particularly at those without specialist
training as map curators but valuable also for those with some experience and/or
training as an *aide-mémoire*. An excellent guide to good practice.

**606 International Federation of Library Associations**, Geography and
Maps Section. *Manual of map curatorship* (in press).

**607 Larsgaard, Mary L.** *Map librarianship: an introduction.* Littleton, Colorado:
Libraries Unlimited, 1978. 330pp. (2nd edn. in press.)
The standard North American textbook. Includes sufficient detail on all aspects
of map care to serve as a working manual for any practising map curator. Has
seven chapters on: Selection and acquisition of maps; Map classification; Map
cataloguing and computer applications; Care, storage and repair of maps; Public
relations and reference services in the map library; Administration of a map
library; and Map librarianship—a brief overview. These chapters are followed
by fifteen appendices and a comprehensive bibliography of some 800 items.

**608 Larsgaard, Mary L.** (ed.). 'Map librarianship and map collections'.
(University of Illinois, School of Library and Information Science.) *Library
Trends*, **29**(3): (1981). 193pp.
Nine papers covering automation, cataloguing and classification, users' research
needs, administration and education, as well as surveys of acquisition, usage,
security, and the seventy largest U.S. collections are followed by two general
papers on map care in Australia and New Zealand. The issue is a useful updating
of Post's work [611] with valuable additions covering security and on Australia
and New Zealand.

**609 Map Curators' Group.** *Introduction to map curatorship. Papers of a one-day
training course.* London: British Cartographic Society, Map Curators' Group,
1982. 57pp.
The papers cover procurement, recording and cataloguing, storage and
handling, and reference services. Of these, the last is the most valuable given the
paucity of other material on the topic.

**610 Nichols, Harold.** *Map librarianship.* 2nd edn. London: Clive Bingley;
Hamden, Connecticut: Linnet Books, 1982. 298pp.
This is the only major British text on the care of maps. As with the other works
listed here, it is designed primarily for the non-specialist—in this case the general
librarian who is required to deal with a map collection. The first six chapters
cover acquisition policies and practice and are followed by a chapter on storage,
six on classification and cataloguing, and one on care and preservation of maps.

**611  Post, J. B.** (ed.). 'Map librarianship'. (Philadelphia: Drexel University.) *Drexel Library Quarterly*, **9**(4): (1973). 90pp.
Eight essays by eminent North American map curators on different facets of map care serve as a helpful introduction to the subject.

**612  Ristow, Walter W.** *The emergence of maps in libraries*. Hamden, Connecticut: Linnet Books; London: Mansell Publishing, 1980. 358pp.
A collection of thirty-five papers written by Ristow between 1946 and 1977. The techniques, etc. described have in many cases been superseded, but these essays are the nearest approach to a history of map care that has yet been written.

**613  Wallis, Helen** and **L. Zögner** (eds.). *The map librarian in the modern world: essays in honour of Walter W. Ristow*. Munich: K. G. Saur, 1979. 295pp.
A *Festschrift* of thirteen papers and a bibliography of Ristow's published works. The descriptions of several map collections are among the most useful contributions.

## Visual aids

**614  Special Libraries Association, Geography and Map Division.** *Introduction to map libraries*. 1982. (Available from Anita K. Oser, Hunter Library, Western Carolina University, Cullowhee, North Carolina 28723.)
A slide/tape presentation of eighty slides which runs for twenty minutes. Aimed at librarians and library science students who are especially interested in map care.

**615  University of Minnesota.** *Map library slide/tape instruction program*. 1981. (Available from Mai Treude, Map Library, S76 Wilson Library, 309 19th Avenue South, University of Minnesota, Minneapolis, Minnesota 55455.)
A programme lasting ten minutes and including sixty-three slides that is designed to help users of map collections to identify and locate the available resources.

## Training package

**616  Forget, J.** *Practical documentation*. IPPF Distribution Department, 18–20 Lower Regent Street, London SW1Y 4PW, England.
A ten-module package that provides basic information on all aspects of librarianship, some of which have relevance for map curators.

## Journals

**617  American Library Association, Map and Geography Round Table (MAGERT).** *base line*. 1980–. 6 issues per annum. (Available from J. Walsh; for address, *see* [402].)
Contains news and articles aimed primarily at members of MAGERT but of interest to a wider readership.

**618  Association of Canadian Map Libraries.** *Bulletin*. 4 issues per annum. (Available from National Map Collection, Public Archives of Canada, Ottawa K1A 0N3, Canada.)

Contains articles and news items of interest to map curators in general as well as some information specifically for ACML members.

**619 Australian Map Circle** (formerly Australian Map Curators' Circle). *Globe.* Irregular publication but usually 1 or 2 issues per annum. (Available from P.O. Box E133, Queen Victoria Terrace, ACT 2600, Australia.)
The bias is towards articles directly relevant to map curators in Australia and New Zealand but with broader applications to all in the profession.

**620 Australian Map Circle.** *Newsletter.* Irregular publication but usually appears between issues of *Globe.* (Available from P.O. Box E133, Queen Victoria Terrace, ACT 2600, Australia.)
Newsletter for members of the AMC which contains information that cannot be delayed until the next issue of *Globe.*

**621 British Cartographic Society.** Map Curators' Group. *Cartographiti.* 1982–. 4 issues per annum. (Available from C. Perkins, 9 Kiln Lane, Hadfield, Hyde, Cheshire SK14 7AU, England.)
This newsletter is the only British publication produced solely for map curators, although full-length articles occasionally appear in the parent society's journal (the *Cartographic Journal*) or in the *Bulletin* of the Society of University Cartographers).

**622 New Zealand Map Keepers' Circle.** *Newsletter.* 1976–. 2 issues per annum. (Available from P. L. Barton, P.O. Box 10–179, Terrace Post Office, Wellington, New Zealand.)
Features articles as well as news items of current events, etc. for NZMKC members.

**623 Southern African Map Collectors' Association/Suid Afrikaanse Kartversamelaarsvereniging.** *The Map Reader.* 1981–. 2 issues per annum. (Available from University Library, University of Natal, P.O. Box 375, Pietermaritzburg 3200, South Africa.)
Contains news items and articles on map care, history of cartography and map collecting.

**624 Special Libraries Association, Geography and Map Division.** *Bulletin.* 1947–. 4 issues per annum. (Available from Special Libraries Association, 9927 Edward Avenue, Bethesda, Maryland 20014.)
The oldest journal for map curators. The *Bulletin* serves as a medium of exchange of information, news and research in the field of geographic and cartographic bibliography, literature and libraries. It includes articles on research problems, technical services and other aspects of cartographic and geographic literature, libraries and collections. A major source of information and an absolutely indispensable aid to all libraries and map curators. Also includes much information of value to cartographers, geographers and all whose work or interests are concerned with maps of any kind.

**625 Western Association of Map Libraries.** *Information Bulletin.* 1968–. 3 issues per annum. (Available from S. D. Stevens, University Library, University of California, Santa Cruz, California 95064.)
Although concentrating on the needs of the North American West, this journal has a much wider catchment for both news and articles than might be expected.

**626 Western Association of Map Libraries.** *Occasional Papers.* 1973–. Irregular publication. (Available from S. D. Stevens, University Library, University of California, Santa Cruz, California 95064.)
A series of monographs mostly consisting of carto-bibliographies and union lists.

## Map acquisition

### Bibliography

**627 Allison, Brent.** 'Map acquisition: an annotated bibliography'. Santa Cruz, California: Western Association of Map Libraries, *Information Bulletin*, **15**(1): 16–25 (1983).

### Major work

**628 Wise, Donald A.** 'Cartographic sources and procurement problems'. New York: Special Libraries Association, *Special Libraries*, No. 68: 198–205 (1977). With appendices published separately in Special Libraries Association, Geography and Map Division, *Bulletin*, **112**: 19–26 (1978); **113**: 65–68 (1978); **114**: 40–44 (1978); **115**: 35–50 (1979).
Apart from the established textbooks this is the only survey of the problems of acquisition. The appendices are particularly valuable listings of sources of both governmental and commercial mapping.

### Source material

**629 Geo Center Internationales Landkartenhaus.** *Geokatalog Band 2.* Stuttgart: Geo Center Internationales Landkartenhaus, 1975–83. *c.* 1600pp.
The trade catalogue of the premier European map supplier; the regular updating service, GEOKARTEN BRIEF, makes this the nearest thing to an international carto-bibliography in existence.

**630 Winch, Kenneth L.** *International maps and atlases in print.* 2nd edn. London and New York: R. R. Bowker, 1976. 966pp.
Developed from the trade catalogue of the map retailer, Edward Stanford Ltd., this international carto-bibliography is now becoming a little dated and regrettably does not have an updating service.

**631 York University Library.** *Map sources directory.* 3rd revised edn. Downsview, Ontario: Map Library, York University, 1983.
Gives updated listings of commercial and governmental publishers. Especially useful for North American governmental agencies (federal and state/provincial).

## Dealers in antiquarian maps

**632 Ritzlin, George.** *World directory of dealers in antiquarian maps.* Chicago: Chicago Map Society, 1977. 28pp.
This directory was compiled with the help of the staff of the Hermon Dunlap Smith Center for the History of Cartography at the Newberry Library.

**633** *The Map Collector.* Church Square, 48 High Street, Tring, Hertfordshire HP23 5BH, England.
Issues a continuing directory of dealers, probably the most comprehensive to be found in any source.

**634** *The ABMR map of antiquarian and secondhand bookshops in Central London.* Compiled by Peter Stockham. (Available from ABMR Publications Ltd., 52 St. Clements Street, Oxford OX4 1AG, England.)

# Map classification and cataloguing

## Bibliography

**635 Sadler, Judith De B.** 'The organization of maps: a bibliography'. New York: Special Libraries Association, Geography and Map Division, *Bulletin*, **126**: 13–18 (1981).
An important list of eighty-five items including automated systems. The article is complementary to that of Schorr [601].

## Major works

**636 Chan, Lois M.** *Library of Congress subject headings; principles and application.* Littleton, Colorado: Libraries Unlimited, 1978.

**637 Gorman, Michael** and **P.W. Winkler** (eds.). *Anglo-American cataloguing rules (AACRII).* London: Library Association; Ottawa, Canadian Library Association; Chicago, American Library Association, 1978.

**638 International Federation of Library Associations and Institutions.** Joint Working Group on the International Standard Bibliographic Description for Cartographic Materials. *Cartographic materials in UNIMARC: the proposals of a sub-group of the Joint Working Group on ISBD (CM).* London: IFLA International Office for UBC, 1979. 18pp.

**639 International Federation of Library Associations and Institutions.** Joint Working Group on the International Standard Bibliographic Description for Cartographic Materials. *ISBD (CM): International Standard Bibliographic Description for cartographic materials recommended by the Joint Working Group on ISBD (CM)* London: IFLA International Office for UBC, 1977. 68pp.

**640 International Federation of Library Associations and Institutions.** Working Group on Contents Designators. *UNIMARC: the universal format for the* *exchange of bibliographic records among national libraries and bibliographic agencies.* 2nd edn. London: IFLA International Office for UBC, 1980.

**641 Lea, Graham, James Shearer** and **Donald Paterson.** 'Computerized indexing of the Institute of Geological Sciences (UK) geological map collection'. New York: Special Libraries Association, Geography and Map Division, *Bulletin,* **112**: 27–46 (1978).

**642 Libault, A.** (ed.). *Final report on the classification of geographical books and maps.* Bad Godesberg: Institut für Landeskunde, 1964. 210pp.
Discusses the UDC system and provides tables of classification numbers.

**643 Library of Congress, Subject Cataloguing Division.** *Library of Congress Classification, Class G: Geography, Maps, Anthropology, Recreation.* 4th edn. Washington, D.C.: Library of Congress, 1976.

**644 Library of Congress, Subject Cataloguing Division, Processing Services.** *Library of Congress Subject Headings.* 9th edn. Washington, D.C.: Library of Congress, 1980.

**645 Merrett, C. E.** *Map cataloguing and classification: a comparison of approaches.* Sheffield: Postgraduate School of Librarianship and Information Science. Occasional Paper No. 7, 1976. 30pp.

**646 Ministry of Defence, Chief of the General Staff.** *Manual of Map Library Classification and Cataloguing (GSGS 5307).* London: Ministry of Defence, 1978. 282pp.

**647 Moore, Barbara N.** *A manual of AACRII examples for cartographic materials.* Lake Crystal, Minnesota: Soldier Creek Press, 1981.

**648 Selmer, Marsha L.** 'Map cataloguing and classification methods; a historical survey'. New York: Special Libraries Association, Geography and Map Division, *Bulletin,* **103**: 7–12 (1976).

**649 Shepherd, Ifan** and **Steve Chilton.** 'MAPLIB: an automated map enquiry system'. *SUC Bulletin,* **14**(2): 1–23 (1980).

**650 Stibbe, Hugo L. P.** (ed.). *Cartographic materials: a manual of interpretation for AACRII.* London: Library Association; Ottawa: Canadian Library Association; Chicago: American Library Association, 1982. 258pp.

**651 Stibbe, Hugo L. P.** *MARC maps, the history of its development and current assessment.* Utrecht: Geografisk Instituut van de Rijksuniversiteit, Organisation for Information Policy, 1976. 293pp.

**652 Webster, Graham.** 'Cartographic materials and MARC'. London: British Cartographic Society. *The Cartographic Journal,* **19**(1): 60–67 (1982).

# Storage and conservation

*Bibliography*

**653 Perry, Joanne M.** 'Map storage methods: a bibliography'. New York: Special Libraries Association, Geography and Map Division, *Bulletin*, **131**: 14–15 (1983).

This list of twenty-one items represents the references to 'Vertical map storage' from *Special Libraries* (New York: Special Libraries Association), **73**(3): 207–212 (1982).

*Major works*

**654 Kidd, Betty.** 'Preventative conservation for map collections'. (New York: Special Libraries Association) *Special Libraries*, **71** (12): 529–538 (1980).

**655 LeGear, Clara Egli.** *Maps: their care, repair and preservation in libraries.* Washington, D.C.: Map Division, Reference Department, Library of Congress, 1956. 75pp.

Contains invaluable advice for libraries and curators of maps on a wide range of topics, among them the following: cataloguing and classification; filing; folding and dissecting; storage; lamination; mounting of maps. Globes, relief models and atlases are dealt with as well as maps, and advice is offered on mounting exhibitions and the layout of a map room. There is a bibliography of over 130 items.

**656 Royal Geographical Society.** *The storage and conservation of maps.* London: Royal Geographical Society, 1955. 8pp.

A report prepared by a committee of the Royal Geographical Society in 1954, the terms of reference being 'to draw up ideal standards of map conservation for the guidance of map curators'. The committee was assisted by replies to a questionnaire sent out to libraries and map collections. Includes a short bibliography of fifteen items. Inevitably now rather dated.

**657 Wardle, D. B.** *Document repair.* London: Society of Archivists, 1971. 84pp. (Society of Archivists Handbooks vol. 1.) (Available from G. F. Osborn, Room 74, Public Library, Marylebone Road, London NW1 5PS, England.)

Includes a useful section on the repair of maps and large documents.

**658 Ehrenberg, Ralph E.** *Archives and manuscripts: maps and architectural drawings.* Chicago: Society of American Archivists, 1982. 64pp.

A manual designed for archivists who have no special background in maps but work with them. Divided into sections on accessions and appraisal; arrangement; description; conservation; storage; reference; access. Includes a glossary, a bibliography and a list of equipment suppliers.

*Visual aids*

**659 National Audio-Visual Center.** *Storage and care of maps.* Washington, D.C.: National Audio-Visual Center, 1980.

## Reader services

### Major works

**660 Cobb, David.** 'Reference service; map librarianship's forgotten topic'. Map Curators' Group. *Introduction to map curatorship. Papers of a one-day training course*, pp. 17–28. London: British Cartographic Society, Map Curators' Group, 1982.
Includes references to fourteen books and papers. This paper is the most recent survey of a subject that has received little attention.

**661 Treude, Mai.** 'Reference services with maps'. New York: Special Libraries Association, Geography and Map Division, *Bulletin*, **102**: 24–29 (1975).
As with Cobb [660] and Winearls, Batchelder and Woodward [662], the appearance of such a short article under the heading of 'major works' indicates the lack of any more definitive study.

**662 Winearls, Joan, Bob Batchelder** and **Frances Woodward.** 'Orientation of users to the map collection'. Ottawa: Association of Canadian Libraries, *Bulletin*, **41**: 25–41 (1980).

### Visual aids

**663 University of Minnesota.** *Map library slide/tape instruction program*
*See* [615].

## Map Collections

### Directories

### World

**664 Ristow, Walter W.** (ed.). *World directory of map collections*. Compiled by the Geography and Map Libraries Sub-Section. Munich: Verlag Dokumentation, 1976. 326pp.
Details 285 map collections in forty-five countries. As well as providing information on holdings of maps, atlases, gazetteers, aerial photographs and other reference material it lists staff, size and type of collection, reproduction facilities, reference facilities, etc.

### Australia

**665 Rauchle, Nancy M.** *Map collections in Australia: a directory*. 3rd edn. Canberra: National Library of Australia, 1980. 141pp.
Identifies significant Australian map resources and includes information on staff, physical form of holdings and map organization systems.

### Canada

**666 Dubreuil, Lorraine.** *Directory of Canadian map collections*. 4th edn. Ottawa: Association of Canadian Map Libraries, 1980. 144pp.

Information provided includes name of each library, personnel, acquisition policy, regulations and cataloguing.

## France

**667 Briend, A. M.** and **D. Gabay.** *Répertoire des cartothèques de France.* Paris: Laboratoire d'Information et de Documentation en Géographie, 1980–81. 96pp.

## Germany

**668 Zögner, Lothar.** *Verzeichnis der Kartensammlungen in der Bundesrepublik Deutschland und Berlin (West).* Im Auftrag der Staatsbibliothek Preussischer Kulturbesitz mit Unterstützung der Deutschen Forschungsgemeinschaft. Wiesbaden: Harrassowitz, 1983. 450pp.

## Netherlands

**669 Van Slobbe, Annemieke.** *Gid voor kaartenverzamelingen in Nederland.* Amersfoort: Nederlandse Vereniging voor Kartografie; Alphen aan den Rijn: Uitgeverij Canaletto, 1980. 207pp.

## United Kingdom

**670 Bond, Barbara A.** *A directory of U.K. map collections.* British Cartographic Society, 1983. 28pp. (Map Curators' Group Publication No. 1.)
Lists the five map collections of the copyright libraries and then 139 map collections arranged alphabetically. Details are provided of the name and position of the person responsible for the collection, address and telephone number for contact and a description of the major interests of each collection.

## United States

**671 Carrington, David K.** and **R. W. Stephenson.** *Map collections in the United States and Canada: a directory.* 4th edn. New York: Special Libraries Association, Geography and Map Division, Directory Revision Committee, 1985. 192pp.
Entries are arranged alphabetically by city within a state or province. Information is given concerning staff, specializations, reproduction facilities, inter-library loan policy, etc.

**672 Bergen, J. V.** 'Map collections in Midwestern universities and colleges'. *The Professional Geographer,* **XXIV**(3): 245–252 (August 1972).
A survey of campus collections in the Midwest. Illustrates some of the problems of making a complete inventory of map collections.

# Geographical Directory

**673** *Orbis Geographicus 1968–72. World directory of geography, compiled by E. Meynen.* Part 1, *Societies, institutes, agencies.* Wiesbaden: Franz Steiner Verlag, 1970.
Lists mapmaking agencies, cartographic societies and on pp. 288–301 lists 129 major map collections throughout the world, arranged by country, and giving the name of the collection, its address, date of foundation, director of the library

and head of the map division. An invaluable reference aid for all geography departments, teachers, librarians, cartographers and curators of map collections.

## Types of Map

Many types of published thematic map have normally been dealt with in articles published in scientific journals rather than having a whole book devoted to them. Map types are so varied and the number of published papers so multitudinous that the reader can only be guided to a small selection of the published material. Many of the papers themselves include useful lists of references for further reading and reference can also be made to the general works on cartography.

### Agricultural maps

**674 Heynen, William J.** *Agricultural maps in the National Archives of the United States, ca. 1860–1930.* Washington, D.C.: U.S. National Archives and Records Service, 1976. 25pp. (References Information Paper 75.)
This is one of three guides that list the cartographic records of individual groups in the National Archives. Represents useful reference material for agriculturalists, ecologists, economists, geographers and librarians.

**675 Morgan, B. A.** 'Sources of information on maps relevant to the study of agriculture'. *SUC Bulletin*, **10**(1): 1–10 (1976).

**676 Wrathall, J. E.** 'Agricultural land classification maps'. *SUC Bulletin*, **13**(1): 13–18 (1979).
A useful discussion on the development and aims of the Agricultural Land Classification project; for planners at any level, geographers and agriculturalists.

### Communications: aviation maps

**677 Hopkin, V. David** and **Robert M. Taylor.** *Human factors in the design and evaluation of aviation maps.* RAF Institute of Aviation Medicine, 1979. 249pp. (AGARDograph No. 225; not available to the general public.)
The authors examine the factors influencing the design and evaluation of aviation charts: the ergonomics of the confined cockpit, the very short time in which maps have to be read and decisions made, etc. An important contribution to the literature of map evaluation.

**678 Meine, Karl-Heinz.** 'Aviation cartography'. *The Cartographic Journal*, **3**(1) 31–40 (June 1966).
Traces the growth of aviation cartography from its beginnings at the end of the nineteenth century. Includes a valuable list of twenty-seven references.

### Communications: railway maps

**679 Garnett, David.** 'Maps'. In P. B. Whitehouse (general editor), *Railway relics and regalia*, pp. 66–77. London: Country Life, 1975.

A useful narrative account of a rather neglected subject. The many kinds of railway map are discussed, from preliminary surveys, through network maps, to posters and maps on carriage panels. Of interest to the general reader as well as cartographers and railway enthusiasts.

**680 Harley, J. B.** *Maps for the local historian: a guide to the British sources.* London: National Council of Social Service, for the Standing Conference for Local History, 1972.
Dr. Harley discusses railway maps on pp. 48–50 and provides a brief but useful guide to the cartographic evidence for the study of railways in Britain.

**681 Modelski, Andrew M.** (compiler). *Railroad maps of North America: the first hundred years.* Washington, D.C.: Library of Congress, 1984. 186pp.
A good reference source for railway buffs, transport scholars, geographers and historians as well as librarians.

**682 Wardle, D. B.** 'Sources for the history of railways at the Public Record Office'. *The Journal of Transport History,* **2** (1955–56).
Wardle describes the collection of railway maps that forms part of a sizeable collection of maps of communications held by the Public Record Office in London.

## Communications: road maps

**683 Bonacker, Wilhelm.** *Bibliographie der Strassenkarten.* Bonn–Bad Godesberg: Kirschbaum, 1973. 240pp.
A comprehensive bibliography of road maps with some 4,600 items. The arrangement by continent and then by country. An important reference tool for map curators.

**684 Galneder, Mary.** 'Highway maps and tourist offices: a source list'. Special Libraries Association, Geography and Map Division, *Bulletin,* **92**: 16–21, 34 (June 1973).
A useful list of organizations issuing highway maps in the United States and Canada. Helpful not only to motorists but also to librarians and map curators.

**685 Morrison, Alastair.** 'Experimental maps of road travel speed'. *The Cartographic Journal,* **8**(2): 115–132 (December 1971).
One of a number of articles on road maps that Morrison has had published in *The Cartographic Journal.* This example is a major article describing the ideas behind road travel speed maps and the problems involved in their design. Aimed primarily at practising cartographers.

**686 Nicholson, T. R.** *Wheels on the road. Road maps of Britain, 1870–1940.* Norwich: Geo Books, 1983. 101pp.
Begins with a narrative history of road maps from the early nineteenth century. Includes much information on cycling maps as well as motoring maps. Provides a selective bibliography and a selective carto-bibliography of 195 items (pp.

89–101). For carto-historians, map collectors, cyclists, motorists and the general reader.

## Fire insurance maps

Fire insurance plans developed from the requirements of the fire insurance underwriters and are an important source of land use information. In North America such plans were produced by the Sanborn Company and in Britain their production was dominated by Chas. E. Goad Ltd.

**687** *Fire insurance maps in the Library of Congress. Plans of North American cities and towns produced by the Sanborn Map Company. A checklist compiled by the Reference and Bibliography Section, Geography and Map Division.* Washington, D.C.: Library of Congress, 1981. 773pp.
An important reference aid for librarians and urban geographers and historians.

**688 Hayward, Robert J.** *Fire insurance plans in the National Map Collection.* Ottawa: Public Archives of Canada, National Map Collection, 1974. 171pp.

**689 Rees, Gary W.** and **Mary Hoeber.** *Catalogue of Sanborn atlases at California State University, Northridge.* Santa Cruz, California: Western Association of Map Libraries, 1973. 122pp.

**690 Ristow, Walter W.** 'U.S. fire insurance maps, 1852–1968'. *Surveying and Mapping*, **30**(1): 30 (March 1970).

**691 Rowley, Gwyn.** 'British fire insurance plans: cartography at work'. *SUC Bulletin*, **18**(1): 1–8 (1984).

**692 Rowley, Gwyn.** *British fire insurance plans.* Old Hatfield, Hertfordshire: Chas. E. Goad Ltd., 1984.
Of interest to archaeologists, historians, urban geographers, planners, railway enthusiasts, librarians.

**693 Wrigley, R. L.** 'The Sanborn map as a source of land use information for city planners'. *Land Economics*, **25**: 216–219 (1949).

## Map use in genealogical research

**694 Kidd, Betty H.** *Using maps in tracing your family history.* Ottawa: Ontario Genealogical Society, Ottawa Branch (P.O. Box 8346, Ottawa, Ontario K1G BH8), 1974. 48pp.
Designed primarily for Canadian readers but of interest to genealogists elsewhere.

## Geological mapping

**695 Bailey, E.** *Geological Survey of Great Britain.* London: T. Murphy, 1952. 278pp.
An introductory account of Britain's official geological maps.

**696 Barnes, John W.** *Basic geological mapping.* Milton Keynes: Open University Press, 1981. 112pp. (Geological Society Handbook.)
An introduction to geological mapping for geographers and students of the geosciences.

**697  Boud, R. C.** 'Geological cartography in the undergraduate curriculum'. *The Cartographic Journal*, **8**(2): 159–167 (December 1971).
Based on the teaching of geological mapping in the Department of Earth Sciences, University of Leeds.

**698 Fleet, J. S.** *The first hundred years of the Geological Survey of Great Britain.* London: HMSO, 1937. 280pp.

**699 Kline, Nancy M.** 'Catalogs of state geological survey publications: a source list'. Special Libraries Association, Geography and Map Division, *Bulletin*, **98**: 55–58 (December 1974).
A useful reference aid to the publications and maps of individual states in the United States.

**700 Lea, G., J. Shearer** and **D. Paterson.** 'Computerized indexing of the Institute of Geological Sciences (U.K.) Geological Map Collection'. Special Libraries Association, Geography and Map Division, *Bulletin*, **112**: 27–50 (June 1978).
Discusses development of the collection and its present state before going into processing procedures. Important for map curators and librarians.

**701 Linton, D. L.** 'The ideal geological map'. *Advancement of Science*, **5**(18): 141–149 (1948).

**702 Lobeck, A. K.** *Block diagrams and other graphic methods used in geology and geography.* Amherst, Massachusetts: Emerson-Trussell, 1958.
The classic work on block diagrams. Important for geographers and geomorphologists as well as for geologists.

**703 Robertson, T.** 'The presentation of geological information in maps'. *Advancement of Science*, **13**(50): 31–41 (1956).

## Journal

**704** *Geotimes*; available from Geotimes, 5205 Leesburg Pike, Falls Church, Virginia 22041.
Robert A. Bier and Janice T. Fitzpatrick list important maps received by the U.S. Geological Survey Library—maps from many sources all over the world.

# Geomorphological mapping; terrain maps; relief representation

## Bibliographies

**705 Brandes, Donald.** 'Sources for relief representation techniques'. *The Cartographic Journal*, **20**(2): 87–94 (December 1983).
A list of forty-eight references on landform and slope mapping is included. The article provides a valuable source for geographers, cartographers, geomorphologists and geologists.

**706 Allen, Daniel.** 'Geomorphological maps of Canada: a bibliography of Canadian federal government maps'. Special Libraries Association, Geography and Map Division, *Bulletin*, **90**: 25–43 (December 1972).
A useful reference for map curators, students of geomorphology and others with Canada as a special study area.

**707 Ristow, Walter W.** (compiler). *Three-dimensional maps: an annotated list of references related to the construction and use of terrain models.* 2nd edn. Washington, D.C.: Library of Congress, Map Division, Reference Department, 1964.
Ristow lists works that discuss and describe both the civil and military applications of three-dimensional maps. He includes relief models such as those used in planning, flood control, soil conservation, mining, transportation and recreation.

## Selected works and papers

**708 Curran, J. P.** 'Cartographic relief portrayal'. *The Cartographer*, **4**(1): 28–32 (1967).
Aimed primarily at professional mapmakers, but could also be helpful for geographers and graphic artists.

**709 Fairbairn, David.** 'Gerippe-lines and the representation of relief'. *SUC Bulletin*, **15**(2): 1–5 (1982).
Gerippe-lines are defined as symbols giving a generalized delineation of watersheds and breaks of slope. For geographers, geomorphologists, cartographers and geo-scientists.

**710 Imhof, Eduard.** *Cartographic relief representation.* Berlin: Walter de Gruyter, 1982. 389pp.
The long-awaited translation of Imhof's famous work [711]. While it is primarily intended for cartographers there is much useful material for geographers, geo-scientists and surveyors.

**711 Imhof, Eduard.** *Kartographische Geländedarstellung.* Berlin: Walter de Gruyter, 1965. 425pp.
The major work on terrain representation by the Swiss master-cartographer. Has a valuable bibliography on pp. 405–419.

# 180    ANNOTATED BIBLIOGRAPHY: CONTEMPORARY CARTOGRAPHY

**712 Imhof, Eduard.** *Gelände und Karte.* Erlenbach–Zürich: Eugen Rentsch, 1968. 280pp.
Definitive work by the doyen of Swiss cartographers. Well-illustrated with sketches, map sections and diagrams. A valuable text for cartographers, professional or otherwise.

**713 Keates, J. S.** 'Techniques of relief representation'. *Surveying and Mapping,* **21**(3): 459–463 (1961).

**714 Miller, O. M.** and **C. H. Summerson.** 'Slope zone maps'. *Geographical Review,* **50**(2): 194–202 (1960).

**715 Robinson, A. H.** 'A method for producing shaded relief from aerial slope data'. *Surveying and Mapping,* **8** (1948).

**716 Yoeli, Pinhas.** 'Digital terrain models and their cartographic and cartometric utilisation'. *The Cartographic Journal,* **20**(1): 17–25 (June 1983).
For professional cartographers and geomorphologists.

## Guidebook maps

**717 Otness, Harold M.** 'Guidebook maps'. Special Libraries Association, Geography and Maps Division, *Bulletin,* **88**: 17–23 (1972).
Discusses some of the more popular series of guidebooks and the maps they contain. Of use to librarians, map curators, cartographers and urban geographers.

**718 Otness, Harold M.** *Index to nineteenth century city plans appearing in guidebooks.* Santa Cruz, California: Western Association of Map Libraries, 1980. 84pp.

**719 Otness, Harold M.** *Index to early twentieth century city plans appearing in guidebooks.* Santa Cruz, California: Western Association of Map Libraries, 1978. 91pp.
The two volumes by Otness constitute a unique reference source for librarians.

## Land utilization mapping

**720 Clark, A.** 'The world land use survey'. *Geographica Helvetica,* pp. 27–28 (1976).

**721 Coleman, Alice.** 'The second land use survey: progress and prospect'. *Geographical Journal,* **127**: 168–186 (1961).
Coleman organized the second of Britain's land use surveys and here looks at its progress in the early stages. Mainly for geographers.

**722 Coleman, Alice** and **W. G. V. Balchin.** 'Land use maps'. *The Cartographic Journal,* **16**(2): 97–103 (December 1979).

An important article that summarizes land use mapping in Britain and elsewhere and particularly looks at the role of the second land use survey of Britain. The authors also explore new techniques developed in recent years. There is a useful list of references. A good article for geographers, planners, ecologists, soil scientists, etc.

**723 Coleman, Alice, Janet E. Isbell** and **G. Sinclair.** 'The comparative statics approach to British land-use trends'. *The Cartographic Journal*, **11**(1): 34–41 (June 1974).
Discusses the examination of national land-use trends by a method of comparative statics. A technical article, mainly for cartographers and geographers.

**724 Coleman, Alice.** 'A geographical model for land use analysis'. *Geography*, **54**: 43–55 (1969).

**725 Kostrowicki, J.** 'The Polish detailed survey of land utilization'. *Dokumentacja Geograficzna*, **2** (1964).
A Polish text that looks at methods and techniques of research in mapping land utilization in Poland.

**726 Stamp, L. Dudley.** 'The land utilisation survey of Britain'. *Geographical Journal*, **78**: 40–53 (1931).
The late Professor Stamp discusses the work of the first land use survey of Britain, a survey for which he was the organizer and which was largely based on local geographical surveys carried out by school teachers during the 1920s. For geographers, planners, agriculturalists, etc.

**727 The Architectural Press.** *Land use mapping by local authorities in Britain.* London: The Architectural Press, 1978. 63pp.
A report commissioned by the Department of the Environment and prepared by the Experimental Cartography Unit at the Royal College of Art. It enquires into the difficulties experienced by local authorities in respect of land use information. The report is biased towards mapmaking by computer graphic methods. Of particular interest to local authorities officials and planners but important also for professional cartographers and geographers.

## Maps in literature

**728 Muehrcke, Phillip C.** and **Juliana O. Muehrcke.** 'Maps in literature'. *The Geographical Review*, **LXIV**(3): 317–338 (July 1974).
Maps of imaginary territories have provided the basis for numerous fictional works such as those of J. R. R. Tolkien or R. L. Stevenson's *Treasure Island*. This paper discusses the fascination of maps for literary authors and how they have used them. A reference aid for librarians and full of interest for the general reader.

**729 Post, J. B.** *An atlas of fantasy.* New revised edn. New York: Ballantine, 1979. 210pp.

J. B. Post of the Free Library of Philadelphia illustrates many maps of imaginary lands that have featured in literature. A reference aid for librarians, students of literature and for cartographers and illustrators.

## Mental maps

**730 Cromley, R., Karl Raitz** and **Richard Ulack.** 'Automated cognitive mapping'. *Cartographica*, **18**(4): 36–50 (1981).
The objectives are twofold: first, to assess the types of cognitive study that might lend themselves to large sample size measurement; and secondly, to provide an automated method that allows analysis of a broad range of cognitive image elements. For cartographers, geographers and psychologists.

**731 Downs, R. M.** and **David Stea.** *Maps in minds: reflections on cognitive mapping.* New York: Harper & Row, 1977. 284pp.
A geographer and a psychologist explore aspects of cognitive mapping, which they define as a reflection of the world as some person believes it to be. Their studies are largely drawn from experiences in the United States of America.

**732 Downs, R. M.** and **David Stea.** *Image and environment: cognitive mapping and spatial behaviour.* Chicago: Aldine, 1973. 439pp.
A collection of essays that investigate the conceptions held by different persons of their place in space. Recommended to geographers, psychologists and urban planners.

**733 Gould, Peter** and **Rodney White.** *Mental maps.* Harmondsworth, Middlesex: Penguin Books, 1974. 204pp.
A study of the geography of 'perception' and of the mental images we form of places which are important in making decisions as to where we live, take holidays, site new factories and so on. Gould and White survey mental maps at work in America, Sweden, Asia, Africa and Great Britain and analyse the implications of their findings. Important to all concerned with the future planning of the environment. Has a useful list of references.

## Oceanographic mapping

**734 Kerr, Adam J.** (ed.). *The dynamics of oceanic cartography.* Toronto: University of Toronto Press, summer 1980. 181pp. (Cartographica Monograph 25, vol. 17, No. 2.)
This important collection of papers is divided into three parts:
Part I. Map projections in marine science; the illustration of oceanic data, scalars; the illustration of oceanic data, vectors; the use of computer graphics; the impact of different mathematical approaches to contouring.
Part II. Oceanographic and marine mapping: a decade of experience at the Experimental Cartography Unit, London.
Part III. Study of some cartographic techniques used by British, American and Japanese oceanographers.
References at ends of each chapter. Technical material for oceanographers, hydrographers and cartographers.

**735 Linton, R.** (compiler). *Methods of display of ocean survey data.* Swindon: National Environmental Research Council, 1983.

**736 Richardson, A.** 'A study of some cartographic techniques used by British oceanographers'. *The International Hydrographic Review,* **LII**(1): 103–112 (January 1975).

## Panoramic maps

**737 Berann, Heinrich C.** and **Heinz A. Graefe.** *Die Alpen im Panorama.* Frankfurt am Main: Verlag Weidlich, 1966. 206pp.
A superbly illustrated volume, with German text, of the Alpine panoramic maps by the Austrian artist Heinrich Berann. For cartographers, graphic artists, advertisers and the general reader.

**738 Berann, Heinrich C.** *Berann's panoramas: wonderful world of Alps and Himalayas.* Tokyo: Jitsugyo No Nihon Sha, 1980.

**739 Garfield, T.** 'The Panorama and Reliefkarte of Heinrich Berann'. *SUC Bulletin,* **4**(2): 52–61 (summer 1970).
Discusses the making of Berann's superb panoramas.

**740 Wood, Michael.** 'The panoramic map of central Scotland'. *SUC Bulletin,* **12**(1): 1–7 (1984).
Discusses each stage in the process of producing a panoramic map of a highland region. For cartographers, graphic artists, tourism information officials and the general reader.

## Photomaps and orthophotomaps

**741 Hill, A. R.** *Cartographic performance: an evaluation of orthophotomaps.* London: Royal College of Art, Experimental Cartography Unit, 1974. 238pp.
Evaluates six graphic forms of photomap. Technical material, primarily for professional cartographers.

**742 Thrower, N. J. W.** and **J. R. Jenson.** 'The orthophoto and orthophotomap: characteristics, development and application'. *The American Cartographer,* **3**(1): 39–56 (April 1976).

## Population maps; ethnographic maps

**743 Barbour, K. M.** 'Population mapping in Sudan'. In K. M. Barbour and R. M. Prothero (eds.), *Essays on African population,* pp. 63–81. London: Routledge & Kegan Paul, 1961.
Discusses the production of population maps of Sudan on the basis of post-war censuses. For demographers, sociologists, anthropologists and geographers.

**744 Dixon, O. M.** 'Methods and progress in choropleth mapping of population density'. *The Cartographic Journal*, **9**(1): 19–29 (June 1972).
Dixon reviews the methods and weakness of choropleth mapping with special relevance to population density. For geographers, cartographers, demographers and sociologists.

**745 Gordon, M.** 'Cartography for census purposes'. *World Cartography*, **13** (1975).

**746 Hilton, T. E.** 'Population mapping in Ghana'. In K. M. Barbour and R. M. Prothero (eds.), *Essays on African population*, pp. 83–95. London: Routledge & Kegan Paul, 1961.
Discusses the making of population maps of Ghana based on post-war censuses.

**747 Hunt, A. J.** and **H. A. Moisley.** 'Population mapping urban areas'. *Geography*, **35**(1): 79–89 (1960).

**748 Hunt, A. J.** 'Urban population maps'. *Town Planning Review*, **23**(3): 238–248 (1953).

**749 Michigan Ethnic Heritage Studies Center.** *Survey and ethnic mapping procedure*. Detroit, Michigan: June 1975. 36pp.
Describes the rationale and procedures for surveying and mapping the ethnic composition of local units in Detroit and its suburbs. For demographers, sociologists, anthropologists, geographers and cartographers.

**750 Prothero, R. M.** 'Population maps and mapping in Africa, south of the Sahara'. In K. M. Barbour and R. M. Prothero (eds.), *Essays on African population*, pp. 63–81. London: Routledge & Kegan Paul, 1961.
A general survey of the work that has been done. For geographers and demographers.

**751 Prothero, R. M.** 'Problems of population mapping in an underdeveloped territory (Northern Nigeria)'. *Nigerian Geographical Journal*, **III**: 1–7 (1960).

**752 Seavey, Charles A.** '1980 census geography and maps'. Special Libraries Association, Geography and Map Division, *Bulletin*, **129**: 10–26 (September 1982).

**753 Storrie, M. C.** and **C. I. Jackson.** 'A comparison of some methods of mapping census data of the British Isles'. *The Cartographic Journal*, **4**(1): 38–43 (1967).

**754 William-Olsson, W.** 'The Commission on a world population map: history, activities and recommendations'. *Geografiska Annaler*, **45**(4): 243–250 (1963).

## Propaganda maps

**755 Ager, J.** 'Maps and propaganda'. *SUC Bulletin*, **11**(1): 1–15 (1977).
The emphasis is on political propaganda. For cartographers, graphic artists, economists, politicians.

**756 Quam, L. O.** 'Uses of maps in propaganda'. *Journal of Geography*, **42**: 21–32 (1943).
Again the emphasis is on political propaganda.

**757 Soffner, H.** 'War on the visual front: use of maps, charts and diagrams for purposes of propaganda'. *American Scholar*, **11**: 465–476 (1942).

**758 Thomas, L.** 'Maps as instruments of propaganda'. *Surveying and Mapping*, **IX**(2): 75–81 (1949).

**759 Yanker, G.** *Prop art.* London: Studio Vista, 1972. 256pp.
Contains over 1,000 political posters including some examples of different types of propaganda map. Useful reference for graphic artists and advertisers.

## Soil survey

**760 Bartelli, Lendo J.** 'General soil maps—a study of landscapes'. *Journal of Soil and Water Conservation*, **21**: 3–6 (Jan./Feb. 1966).

**761 Proehl, Karl H.** 'The soil survey: an annotated cartographic tool'. Special Libraries Association, Geography and Map Division, *Bulletin*, **115:** 18–25 (March 1979).
Proehl highlights the importance of soil surveys for librarians.

**762 Rothamsted Experimental Station.** *Soil Survey of England and Wales.* A six-page folder describing the work of the Soil Survey available from the Publications Officer, Rothamsted Experimental Station, Harpenden, Hertfordshire AL5 2JQ, England.

**763 White, Leslie Paul.** *Aerial photography and remote sensing for soil survey.* London and New York: Oxford University Press, 1977. 104pp.
For soil scientists, geographers, photogrammetrists and cartographers.

## Tactual mapping—mapping for the blind

**764 James, G. A.** and **J. D. Armstrong.** *Handbook on mobility maps.* (Available from G. A. James, 223 College Street, Long Eaton, Nottinghamshire, England.)

**765 Wiedel, J. W.** 'Tactual maps for the visually handicapped. Some developmental problems'. *The Professional Geographer*, **18**(3): 132–139 (May 1966).
An introduction for geographers, cartographers, social workers, doctors.

**766 Wiedel, J. W.** and **P. A. Groves.** 'Towards a technique for tactual map design and reproduction'. *SUC Bulletin*, **4**(1): 12–18 (December 1969).
Examines the ways of preparing and reproducing tactual maps. Specifically written for professional cartographers but of interest to geographers, social workers and others.

**767 Wiedel, J. W.** and **P. A. Groves.** *Tactual mapping: design, reproduction, reading and interpretation.* College Park, Maryland: Department of Geography, Maryland University, 1969. 129pp.
The final report on research into the design of tactual maps with many illustrations.

## Treasure maps

### Carto-bibliography

**768 Wise, Donald A.** *A descriptive list of treasure maps and charts in the Library of Congress.* Washington, D.C.: Government Printing Office, 1973. 30pp.

### Other work

**769 Wise, Donald A.** 'Treasure maps'. Special Libraries Association, Geography and Map Division, *Bulletin*, **101**: 15–22 (September 1975).
Includes a useful list of references for librarians, map curators, map collectors, the general public.

## Vegetation maps

### Bibliography

**770 Küchler, A. W.** *International bibliography of vegetation maps.* Lawrence, Kansas: University of Kansas Libraries, 1965–70. 4 vols.
A major reference tool for geographers, map curators, librarians, agriculturalists, botanists, economists, etc.

### Other works

**771 Küchler, A. W.** 'Classification and purpose in vegetation maps'. *Geographical Review*, **XLVI**: 155–167 (1956).

**772 Küchler, A. W.** 'Some uses of vegetation mapping'. *Ecology*, **34**: 629–636 (1953).

**773 McGrath, G.** 'Further thoughts on the representation of vegetation on topographic and planimetric maps'. *The Cartographic Journal*, **3**(2): 48–50 (December 1966).
Includes a bibliography. For professional cartographers, geographers, biologists.

**774 Sinclair, G.** 'Mapping semi-natural vegetation'. *The Cartographic Journal*, **18**(1): 48–50 (June 1981).

Discusses three approaches to mapping semi-natural vegetation: remote sensing, direct sensing and ultra-direct sensing. For cartographers and photogrammetrists.

## Weather maps; pollution maps

**775 Kadmon, Naftali.** 'Photographic, polyfocal and polar-diagrammatic mapping of atmospheric pollution'. *The Cartographic Journal*, **20**(2): 121–126 (December 1983).
Discusses suitable graphical methods of representation to display, predict and combat atmospheric pollution. For cartographers and geographers as well as planners.

**776 Menmuir, P.** 'The automatic production of weather maps'. *The Cartographic Journal*, **11**(1): 12–18 (June 1974).
Describes a method for the automatic plotting or display in symbolic form of the weather observations on to weather maps. For cartographers, geographers, meteorologists.

**777 Pedgley, D. E.** *A course in elementary meteorology.* London: HMSO, 1962. 189pp.

**778 Wickham, P. G.** *The practice of weather forecasting.* London: HMSO, 1970. 187pp.

## *Education; Careers in Cartography*

**779 Anson, Roger W.** *Careers in cartography.* Oxford: British Cartographic Society. n.d. 19pp.
A useful booklet for anyone wishing to investigate the possibility of a career in cartography. Lists the major employers in Great Britain; the educational establishments offering courses in cartography; entry requirements for both courses and jobs.

**780 British Cartographic Society.** *Cartographic education for the future.* 1984. (BCS Special Publication No. 3.)
Several contributors review the whole field of cartography courses and their relevance to various governmental and commercial organizations.

**781 Coulson, M. R. C.** (ed.). *The introductory cartography course at Canadian universities.* Ottawa: Canadian Cartographic Association, 1980. 92pp.
Twelve authors describe the courses offered at their universities. Useful for anyone contemplating the teaching of cartography at university level.

**782 Dahlberg, R. E.** 'Cartographic education in U.S. colleges and universities'. *The American Cartographer*, **4**(2): 145–156 (October 1977).

**783 Keates, J. S.** 'Automation and education in cartography'. *The Cartographic Journal*, **11**(1): 53–55 (June 1974).
Discusses how major changes in the systems of map production are bound to affect the content and requirements of educational programmes.

**784 Kent, R. B.** 'Academic geographer/cartographers in the United States: their training and professional activity in cartography'. *The American Cartographer*, **7**(1) 59–66 (April 1980).

## Mapping sciences: a bibliography of educational items

**785 Steward, H.** *Education and training in mapping sciences: a working bibliography.* New York: American Geographical Society, 1969. 72pp.
Steward lists 720 items that cover the fields of cartography, surveying and photogrammetry.

# Ordnance Survey

A great number of books and papers have been written about Great Britain's Ordnance Survey and information can also be gleaned from parliamentary papers, Ordnance Survey Professional Papers, Ordnance Survey Annual Reports, Ordnance Survey leaflets, Parliamentary Reports of Select Committees on the Ordnance Survey and so on. A useful selective bibliography appears on pp. 185–191 of *Ordnance Survey maps: a descriptive manual* [787].

### Major works

**786 Close,** *Colonel Sir* **Charles.** *The early years of the Ordnance Survey.* Originally published in book form, Chatham: The Institution of Royal Engineers, 1926. Reprinted, with an introduction by J. B. Harley, Newton Abbot, Devon: David & Charles, 1969. 164pp.
This was for long the standard history. Deals with the eighteenth-century origins of the Survey and with the development of the English and Irish Surveys up to 1846. Close was Director General of the Ordnance Survey from 1911 to 1922. The reprint is an important reference work with a particularly valuable introduction by J. B. Harley.

**787 Harley, J. B.** *Ordnance Survey maps: a descriptive manual.* Southampton: Ordnance Survey, 1975. 200pp.
A concise picture of the Ordnance Survey's current map production with a description of all the present map series. Contains extracts from maps at different scales, as well as a bibliography. The definitive work for geographers, cartographers, map curators, librarians and map users generally.

**788 Harley, J. B.** and **C. W. Phillips.** *The historian's guide to Ordnance Survey maps.* London: The Standing Conference for Local History, 1964. 51pp.
Written specifically for local historians, this short volume provides a valuable

review of Ordnance Survey production including all the early series and includes index maps for the various series of the one inch to one mile map.

**789 Ordnance Survey.** *A description of Ordnance Survey small scale maps.* Chessington, Surrey: Ordnance Survey Office, 1957. 21pp. 17 plates.

**790 Ordnance Survey.** *A description of Ordnance Survey medium scale maps.* Chessington, Surrey: Ordnance Survey, 1955. 21pp. Plates.

**791 Ordnance Survey.** *A description of Ordnance Survey large scale plans.* Chessington, Surrey: Ordnance Survey, 1954. Plates.
A set of three short volumes illustrated with map extracts which describe all the map series published by the Ordnance Survey up to the mid 1950s.

**792 Seymour, W. A.** (ed.). *A history of the Ordnance Survey.* Folkestone: Wm. Dawson & Sons, 1981. 400pp.
A comprehensive reference work on the development of a body that has been the model for national cartographic organizations throughout the world. An invaluable source of research material for cartographic historians, geographers and all map users.

### Selected references

**793 Aylward, J.** 'The retail distribution of Ordnance Survey maps and plans in the latter half of the nineteenth century—a map seller's view'. *The Cartographic Journal*, **8**(1): 55–58 (1971).

**794 Booth, J. R. S.** *Public boundaries and Ordnance Survey, 1840–1980.* Southampton: Ordnance Survey, 1980. 451pp.
A reference aid for surveyors, local government officials, legal purposes.

**795 Dixon, O. M.** 'The Ordnance Survey 1:50,000 series'. *SUC Bulletin*, **9**(1): 4–9 (1975).
A discussion of the metric series, which broke the tradition of Ordnance Survey one inch to one mile maps.

**796 Gardiner-Hill, R. C.** *The development of digital maps.* Southampton: Ordnance Survey, 1972. 14pp. (Ordnance Survey Professional Papers, New Series, No. 23.)
Outlines the Ordnance Survey approach to cartographic data and computer techniques.

PART IV

# Directory of Organizations

# Organizations Concerned with Cartography

## Map Collections

The following list notes only the major map collections throughout the world. In addition to those establishments mentioned, countless smaller collections are housed in university and public libraries, record offices, local archives and great houses. *A directory of U.K. map collections* (Bond, [670]) has recently been published by the Map Curators' Group of the British Cartographic Society and in it are listed the five main copyright libraries together with 139 map collections throughout the United Kingdom. Similar directories, though few in number, have been published in other countries and Ristow's *World directory of map collections* [664] gives an overall view of the field.

Nichols [610] gives some indication of the type of material to be found in the various types of collection. The enquirer for contemporary local maps, for example, may well find that his needs will be met at his municipal library, though a local branch library would be unlikely to have any reference collection apart perhaps from a few general atlases for student use. The student of local history will probably have to resort to his county record office if he needs to consult early maps, whether printed or manuscript, and the serious student of cartographic development may find what he needs only in the resources of a major map collection such as that of the British Library. Most departments of geography in British and other universities maintain good-sized map collections and are likely to be particularly good for maps of their local area. Limitations of space mean that only a small number of such collections can be cited here—Canada, for example, has a number of important university collections—and it should be stressed once again that the primary purpose of most large collections is to serve accredited readers rather than members of the general public. A letter of enquiry should always be sent as a preliminary approach to the map curator of a particular collection, either seeking advice on a problem related to maps or asking permission to consult material in the collection.

## Major map collections

*Argentina*

**797** Instituto Geográfico Militar, Ave. Cabildo 381, Buenos Aires

*Australia*

**798** National Library of Australia, Canberra

**799** State Library of New South Wales, Macquarie Street, Sydney 2000

*Austria*

**800** Österreichische Nationalbibliothek, Kartensammlung, Josefplatz 1, Vienna I

**801** Österreichisches Staatsarchiv; Abteilung IV, Kriegsarchiv, Stiftgasse 2, Vienna VII

**802** Bibliothek des Bundesamtes für Eich- und Vermessungswesen, Gruppe L [Library of the Federal Survey Office], Krotenthallergasse 3, Vienna VIII

*Belgium*

**803** Bibliothèque Royale de Belgique, 2–4 boulevard de l'Empereur, Brussels 1

*Brazil*

**804** Biblioteca Nacional do Rio de Janeiro, Avenida Rio Branco 219–39, Rio de Janiero

**805** Biblioteca Municipal 'Mário de Andrade', Rua da Consolação 94, São Paulo

**806** Biblioteca e Mapoteca, Instituto Geologico, Avenida Miguel Stefano 3000, CEP 04301, São Paulo

*Bulgaria*

**807** Cyril and Methodius National Library, Boulevard Tolbuhin 11, 1504 Sofia

*Canada*

**808** National Map Collection, Public Archives of Canada, 395 Wellington Street, Ottawa K1A 0N3

**809** Department of Energy, Mines and Resources, Map Library of the Geological Survey of Canada, 601 Booth Street, Ottawa K1A 0E8.

**810** University of British Columbia Library, Map Division, Vancouver, British Columbia V6T 1W5.

## China

**811** National Library of Peking, Peking 7.

## Denmark

**812** Statsbiblioteket i Aarhus [State Library of Aarhus], Universitetsparken, DK-8000 Aarhus C.

**813** Det Kongelige Biblioteks Kort og Billedafdeling [Maps and Prints Department of the Royal Library], 8 Christians Brygge, Copenhagen

**814** Rigsarkivets Kortsamling [Map Collection of the National Archives], 9 Rigsdagsgården, DK-1218 Copenhagen K.

**815** Farvandsdirektoratet, Nautisk Afdeling [Chart Collection of the Naval Department Ministry of Defence], 19 Esplanaden, DK-1263 Copenhagen K.

**816** Geodaetisk Instituts Kortsamling [Map Collection of the Geodetic Institute], Proviantgården, 7 Rigsdagsgården, DK-1218 Copenhagen K.

## Finland

**817** Helsinki University Library (houses the Nordenskiöld Collection), P.O. Box 312, SF-00171 Helsinki 17

## France

**818** Bibliothèque National, Département des Cartes et Plans, 58 rue de Richelieu, Paris 2ᵉ.

**819** Bibliothèque, Centre du Géographie, 191 rue Saint Jacques, F-75005 Paris.

**820** Cartothèque du Laboratoire Intergeo du CNRS, 191 rue Saint Jacques, F-75005 Paris.

## German Democratic Republic

**821** Deutsche Staatsbibliothek, Kartenabteilung, Unter den Linden 8, Berlin W8.

**822** Sächsische Landesbibliothek, Marienallee 12, DDR-806 Dresden

**823** Staatsarchiv Dresden, Archivstrasse 14, DDR-806 Dresden

**824** Kartensammlung des VEB Hermann Haack Geographische-Kartographische Anstalt, Justus-Perthes-Strasse 5–9, DDR-58 Gotha

**825** Deutsche Bücherei, Kartensammlung, Deutscher Platz, DDR-701 Leipzig

## Federal Republic of Germany

**826** Staatsbibliothek Preussischer Kulturbesitz, Kartenabteilung, Potsdamer-strasse 33, D-1000 Berlin 30

**827** Bayerische Staatsbibliothek, Kartensammlung, Ludwigstrasse 16, D-8000 Munich 34

**828** Bayerisches Hauptstaatsarchiv, Abt. Kriegsarchiv, D-8000 Munich 19

## Hungary

**829** Magyar Tudomanyos Akadémia Földrajztudományi Kutató Intezet Könyvtára [Library of the Geographical Research Institute of the Hungarian Academy of Sciences], Budapest.

## India

**830** National Library of India, Maps and Prints Division, Belvedere, Calcutta, 700 027.

## Israel

**831** Map Collection of the National and University Library, Giv'at Ram Campus, Jerusalem.

**832** Micha Grant Library, Geography Department of the Tel-Aviv University, Ramat-Aviv

**833** The Geography Department Map Collection, Hebrew University, Jerusalem

**834** National Maritime Museum, 198 Allenby Road, Haifa

## Italy

**835** Biblioteca Nazional Centrale, Piazza de Cavalleggeri 1A, Florence

**836** Biblioteca della Società Geografica Italiana, Villa Calimontana, Via della Navicella 12, Rome.

**837** Biblioteca Nazionale Centrale Vittorio Emanuelle II, Viale Castro Pretorio, I-00185 Rome.

**838** Biblioteca Nazionale Marciana, Palazzi della Zecca e della Libreria Vecchia, Venice

**839** Museo Correr, Piazza San Marco, Venice

## Japan

**840** Kokuritsu Kokkai Toshokan [National Diet Library], Nagato-cho 1–10–1, Chiyoda-ku, Tokyo 100.

## Mexico

**841** Biblioteca de la Secretaria de Relaciones Exteriores [Library of the Ministry of Foreign Affairs], Ricardo Flores Magón 1, Tlatelolco 3, D.F. Mexico

**842** Biblioteca del Instituto Panamericano de Geografía e Historía, Tacubaya 18, D.F. Mexico.

## Netherlands

**843** Topografische Dienst [Topographical Service], Westvest 9, Delft

**844** Rijksmuseum, Rijksprentenkabinet [State Print-room], Jan Lukenstraat 1a, Amsterdam.

**845** Nederlandsch Historisch Scheepvaartmuseum [Netherlands Historical Maritime Museum], Cornelis Schuytstraat 57, Amsterdam

**846** Universiteitsbibliotheek [University Library], Singel 423, Amsterdam

**847** Algemeen Rijksarchief [General State Archives], Bleyenburg 7, The Hague

**848** Universiteitsbibliotheek [University Library], Rapenburg 70–74, Leiden.

**849** Maritiem Museum 'Prins Hendrik' [Prins Hendrik Maritime Museum], Scheepmakershaven 48, Rotterdam

**850** Rijksuniversiteit, Universiteitsbibliotheek [State University, University Library], Wittevrouwenstraat 11, Utrecht

## New Zealand

**851** Hocken Library, University of Otago, P.O. Box 56, Dunedin

**852** Alexander Turnbull Library, P.O. Box 12–349, Wellington North

## Norway

**853** Utenriksdepartementets Bibliotek [Ministry of Foreign Affairs Library], 7 Juni Plassen, 1 Oslo.

**854** Universitetsbiblioteket [University Library], Bergen.

**855** Det Kongelige Norske Videnskaben Selskab Bibliotek [Royal Norwegian Society of Science and Letters Library], Erling Skakkes Gate 47c, Trondheim.

## Peru

**856** Biblioteca Nacional, Avenida Abancay, Apdo. 2335, Lima

## Philippines

**857** University of the Philippines Library, Quezon City.

**858** Bureau of Coast and Geodetic Survey Library, Manila

**859** National Museum Library, Manila

## Poland

**860** Biblioteka Publiczna M. St. Warszawy [Public Library of Warsaw].

**861** Biblioteka Universytecka w Warszawie [University Library of Warsaw], Krakowskie Przedmieście 26–28, 32, 00.927 Warsaw

**862** Biblioteka Instytutu Geografii i Przestrzennego Zagospodarowania PAN [Library of the Institute of Geography and Spatial Organization of the Polish Academy of Sciences], Krakowskie Przedmieście 30, Warsaw.

**863** Biblioteka Wydziatu Geografii i Studiow Regionalnych, Universytetu Warszawaskiego [Library of the Faculty of Geographical and Regional Studies, University of Warsaw], Krakowskie Przedmieście 30, Warsaw

## Portugal

**864** Biblioteca da Sociedade de Geografia de Lisbõa, Rua des Portas de Santo Antão 100, 1100 Lisbon.

**865** Centro de Estudos de Cartografia Antiga [Centre for Studies in the History of Cartography], Lisbon Section, Rua da Junqueira 86, 1300 Lisbon

**866** Centro de Estudos de Cartografia Antiga [Centre for Studies in the History of Cartography], Coimbra Section, Faculdade de Ciências, Universidade de Coimbra, Coimbra

## Romania

**867** Archivele Statului [State Archives], Bd. Gheorghe Gheorghiu-Dej 29, Bucharest

**868** Biblioteca Centrală de Stat [State Archives Library], Strada Ion Ghica 4, Sector IV, 70018 Bucharest

## South Africa

**869** South African Library, Queen Victoria Street, Cape Town

**870** Johannesburg Public Library, Market Square, Johannesburg

**871** University of Cape Town Library, Rondebosch, 7700 Cape Town

## Spain

**872** Museo Naval de Madrid [Naval Museum], Madrid

**873** Atarazanas Reales [Maritime Museum], Barcelona.

**874** Archivo General de Simancas [Simancas General Archives], Valladolid

*Sweden*
**875** Kungl. Bibliotek [Royal Library], Box 5039, S-10241 Stockholm 5.

**876** Kungl. Vitterhets Historie och Antikvitets Akademien Bibliotek [Library of the Royal Academy of Letters, History and Antiquities], Box 5405, S-114 84 Stockholm.

**877** Universitetsbibliotek [University Library], Lund.

**878** Uppsala Universitetsbibliotek, Uppsala

*Switzerland*
**879** Schweizerische Landesbibliothek [Swiss National Library], 15 Hallwylstrasse, Berne

**880** Staatsarchiv Basel, Martinsgasse 2, CH-4001 Basel

**881** Zentralbibliothek Luzern, Sempacherstrasse 10, CH-6000 Lucerne

**882** Zentralbibliothek Zürich, Zähringerplatz 6, Zürich

**883** Öffentliche Bibliothek der Universität, 20 Schönbeinstrasse, Basel

**884** Bibliothèque Publique et Universitaire de Genève, Promenade des Bastions, CH-1211 Geneva 4

**885** Bibliothèque des Nations Unies, Service Géographique, Palais des Nations, Geneva

*United Kingdom*
**886** Map Library, British Library, Great Russell Street, London WC1B 3DG

**887** Royal Geographical Society, Kensington Gore, London SW7 2AR

**888** Bodleian Library, Oxford OX1 3BG

**889** National Maritime Museum, Greenwich, London SE10 9NF

**890** Cambridge University Library, Cambridge CB3 9DR

**891** Public Record Office, Kew, Richmond, Surrey TW9 4DU

**892** Public Record Office, Chancery Lane, London WC2A 1LR

**893** Guildhall Library, Basinghall Street, London EC2P 2EJ

**894** Map Room of the National Library of Scotland, 137 Causewayside, Edinburgh EH9 1PH

**895** National Library of Wales, Aberystwyth, Dyfed SY23 3BU

*United States of America*

**896** Library of Congress, Geography and Map Division, Washington, D.C. 20540.

**897** U.S. Geological Survey Library, Mail Stop 710, Reston, Virginia 22092

**898** Department of the Interior Library, Washington, D.C. 20240.

**899** Newberry Library, 60 West Walton Street, Chicago, Illinois 60610

**900** New York Public Library, Map Division, 5th Avenue and 42nd Street, New York, NY 10018

**901** Yale University Library, New Haven, Connecticut 06520

**902** Detroit Public Library, 5201 Woodward Avenue, Detroit, Michigan 48202

**903** University of California, UCLA Map Library, Los Angeles, California 90024.

**904** University of Wisconsin, Milwaukee, Library Map Division.

**905** Free Library of Philadelphia, Logan Square, Philadelphia, Pennsylvania 19103.

**906** University of Illinois Library, Map and Geography Library, Urbana, Illinois 61801

**907** National Geographic Society, Cartographic Division Map Library, 17th Street, Washington, D.C. 20036

**908** National Ocean Survey Map Collection, Rockville, Maryland

*U.S.S.R.*

**909** Lenin State Library of the U.S.S.R., Prospekt Kalinina 3, Moscow

**910** Library of the Academy of Sciences of the U.S.S.R., Birzhevaya Liniya 1, Leningrad B-164.

**911** M. E. Saltykov-Shchedrin State Public Library, Sadovaya 18, Leningrad D-69

# Major National Mapping Agencies

Note that in addition to the agencies listed below certain other government departments, such as those responsible for forestry, censuses, etc., produce maps to a greater or lesser extent.

### Afghanistan

**912** Institute of Cartography, Ministry of Mines and Industry, Kabul.

### Algeria

**913** Ministère de l'Industrie et de l'Energie, Direction des Mines et de la Géologie, Service Géologique, Immeuble Maurétania, Agha, Algiers.

**914** Ministère de la Défense Nationale, Institut National de Cartographie, B.P. 32, Hussein-Dey, Algiers.

### Angola

**915** Missão Geografica de Angola, C.P. 432, Huambo.

**916** Servicio de Geologia y Minas, Caixa Postal No. 1260 C, Luanda.

### Argentina

**917** Instituto Geografico Militar, Avda. Cabildo 301, 1426 Buenos Aires.

**918** Instituto Nacional de Geología y Minería, Biblioteca, Perú 562, Buenos Aires

**919** Servicio de Hidrografía Naval, Avenida Montes de Oca 2124, 1277 Buenos Aires

### Australia

**920** Division of National Mapping, Department of National Development and Energy, P.O. Box 548, 7 Morrisett Street, Queanbeyan, New South Wales 2620

**921** Bureau of Mineral Resources, Geology and Geophysics, Department of National Development, P.O. Box 378, Canberra, ACT 2601.

**922** Royal Australian Navy Chart Agency, Hydrographic Service, P.O. Box 1332, North Sydney, New South Wales 2060.

### Austria

**923** Bundesamt für Eich- und Vermessungswesen (Landesaufnahme) in Wien [Section of Land Survey, Topography and Cartography], Krotenthallergasse 3, Vienna VIII.

**924** Geologische Bundesanstalt, Rasumofskygasse 23, A-1031 Vienna III.

### Bahamas

925 Ministry of External Affairs, P.O. Box 792, Nassau.

### Bahrain

926 Topographical Survey Department, Manama, Bahrain.

### Belgium

927 Institut Géographique National (Nationaal Geografisch Instituut), Abbaye de la Cambre 13, B-1050 Brussels.

928 Service Géologique de Belgique, 13 rue Jenner (Parc Leopold), B-1040 Brussels.

### Benin

929 Service Topographique, Ministère des Travaux Public, B.P. 360, Cotonou

930 Mines and Geology Branch, Ministry of Transport and Telecommunications, Box 249, Cotonou

### Bolivia

931 Instituto Geográfico Militar y de Catastro Nacional, Gran Guartel General, Avenue Saavedra, La Paz.

932 Servicio Geológico de Bolivia, Ministerio de Minas y Petroleo, Avenida 16 de Julio, 1769, Casilla Cerreo 2729, La Paz

### Botswana

933 Department of Surveys and Lands, Gaborone

### Brazil

934 Departamento de Documentação e Divulgação, Geográfica e Cartográfica, Instituto Brasileiro de Geográfia, 436 Avenida Beira Mar, Rio de Janeiro

935 Instituto Geográfico e Geológico, São Paulo, S.P.

936 Directoria de Hidrográfia e Navegação, Ministerio de Marinha, Ilha Fiscal, Rio de Janeiro

### Bulgaria

937 Glavno Upravlenie po Geodeziia i Kartografiia [Administration of Geodesy and Cartography], Ul. Alabin 46, Sofia

### Burma

938 Burma Survey Department, 460 Merchant Street, Rangoon

939 Geological Survey, 226 Mahabandola Street, P.O. Box 843, Rangoon

*Burundi*

**940** Ministère de l'Agriculture et de l'Elevage, Affaires Foncières et Cadastres, B.P. 38, Bujumbura

**941** Département de Géologie et Mines, Ministère des Affaires Economiques et Financières, B.P. 745, Bujumbura

*Cameroun (Cameroon)*

**942** Service Géographique du Cameroun, Avenue Mgr. Vogt, B.P. 157, Yaoundé

**943** Direction des Mines et de la Géologie, Ministère des Transports, des Mines, des Postes et des Télécommunications, B.P.70, Yaoundé

*Canada*

**944** Surveys and Mapping Branch, Department of Energy, Mines and Resources, 615 Booth Street, Ottawa, Ontario K1A 0E9.

**945** National Geographical Mapping Division, Department of Energy, Mines and Resources, 580 Booth Street, Ottawa, Ontario K1A 0E9

**946** Canadian Hydrographic Service, 615 Booth Street, Ottawa, Ontario K1A 0A3

**947** Geological Survey of Canada, 601 Booth Street, Ottawa, Ontario

*Central African Republic*

**948** Institut Géographique National, B.P. 165, Bangui

*Chad*

**949** Service des Mines, Département des Travaux Publiques, Fort Lamy

*Chile*

**950** Instituto Geográfico Militar, Departamento de Ventas, Calle Dieciocho 354, Santiago.

**951** Instituto de Investigaciónes Geológica, Agustinas 785, 5° Pisa, Casilla 10465, Santiago.

**952** Instituto Hidrográfico de la Armada, Casilla 324, Valparaiso

*China*

**953** National Bureau of Surveying and Mapping, Baiwanzhuang, Beijing

**954** National Geological Survey, 942 Chukiang Road, Nanking, Kiangsu

*Colombia*

**955** Instituto Geográfico 'Agustin Codazzi', Departamento Cartográfico, Avenida Ciudad de Quito 48–51, Bogotá

**956** Instituto Nacional de Investigaciónes Geológico Mineras, Carrera 30, No. 51–59m Bogotá

*Congo*

**957** Institut Géographique National, Centre en Afrique Equatoriale, B.P. 125, Brazzaville

*Costa Rica*

**958** Instituto Geográfico Nacional, Apartado 2272, San Jose

**959** Geological Survey, University of Costa Rica, Department of Geology, San Pedro de Montes de Oca, San Jose

*Cuba*

**960** Técnico de Cartografía, Instituto Cubano de Geodesía y Cartografía, Loma Y39, Nuevo Vedado, Havana

*Cyprus*

**961** Drawing Office, Department of Lands and Surveys, Ministry of the Interior, Nicosia

**962** Geological Survey Department, 3 Electra Street, P.O. Box 809, Nicosia.

*Czechoslovakia*

**963** Ustredni Sprava Geodezie a Kartografie [Central Office for Geodesy and Cartography], Hybernska 2, Prague 1.

**964** Kartografické Nakladelstoi [Cartographic Publishing House], Frantiska Krizka, Prague 7.

**965** Ustredni ustav Geologicky [Central Office for Geology], Nostel Radu Republiky a Radu Prace, Malostranske Nam 19, Prague 1.

*Denmark*

**966** Geodaetisk Instutut [Geodetic Institute], Rigsdagsgården 7, DK-1218 Copenhagen.

**967** Danmarks Geologiske Undersøgeke, Copenhagen.

**968** Farvandskirektorat Nautisk Afdeling [nautical charts], Esplanaden 19, DK-1263 Copenhagen

*Dominican Republic*

**969** Consejo National de Geografía y Cartografía, Ciudad Trujillo

*Eduador*

**970** Instituto Geografía Militar, Ejército Ecuatoriano, Apt. 2435, Quito

**971** Director General de Geología y Minas, Ministerio de Recursos Naturales y Energéticos, Apt. 23-A, Quito

*Egypt*

**972** Survey Department, Al-Harom Street, Aerial Survey Building, Giza, Cairo

**973** Geological Survey and Mining Authority, 3 Salah Salem Street, Abbasia, Cairo

*El Salvador*

**974** Ministerio de Obras Públicas, Instituto Geográfico Nacional [National Geographical Institute], Avenida Juan Bertis 79, San Salvador

*Ethiopia*

**975** Government Mapping and Geography Institute, P.O. Box 597, Addis Ababa

**976** Ministry of Mines, P.O. Box 486, Addis Ababa

*Fiji*

**977** Ministry of Lands and Mineral Resources, Technical Section, Government Building, P.O. Box 2222, Southern Cross Road, Suva

*Finland*

**978** Maanmittaushallitus (Map Service of the National Board of Survey), Etelaesplanadi 10, SF-00130 Helsinki 13

**979** Geological Survey of Finland, Helsinki

**980** Merenkulkuhallitus Merikarttaosasto (Hydrographic Department, Board of Navigation), P.O. Box 158, SF-00141 Helsinki 14

*France*

**981** Institut Géographique National, 136 bis rue de Grenelle, Paris 7ᵉ.

**982** Service du Cadastre, 6 rue des Pyramides, F-75001 Paris

**983** Service Géologique National, Bureau de Recherches Géologiques et Minières, B.P. 6009, Avenue de Concyr, F-45060 Orléans

**984** Institut National de Recherche Agronomique, Service du Carte des Sols, Orléans

**985** Service Hydrographique et Océanographique de la Marine, 3 avenue Octave Greard, F-75200 Paris

**986** Centre d'Etudes et de Réalisations Cartographiques et Géographiques, Centre National de la Recherche Scientifique, 191 rue Saint-Jacques, F-75005 Paris

*French Guiana*

**987** Bureau des Recherches Géologiques et Minières, B.P. 42, Cayenne

*Gabon*

**988** Institut Géographique National, Agence de Libreville, B.P. 13108, Libreville

*German Democratic Republic*

**989** Verwaltung Vermessungs- und Kartenwesen, Mauerstrasse 29–32, D-1086 Berlin

*Federal Republic of Germany*

**990** Institut für Angewandte Geodäsie [Institute of Applied Geodesy], Richard Strauss Allee 11, D-6000 Frankfurt am Main 70

**991** Bundesanstalt für Bodenforschung, Bad Godesberg

**992** Deutsches Hydrographisches Institut, Bernhard Necht Str. 78, D-2000 Hamburg 4.

N.B. Each of the *Länder* has its own survey department.

*Ghana*

**993** Survey Department, P.O. Box 191, Cantonments, Accra

*Greece*

**994** Topographic Service, Likourgon 12, Athens

**995** Institute of Geological and Mining Research, 70 Messoghion Street, Athens 608

**996** Hellenic Hydrographic Service, B.S.T. 902, Athens

*Guadeloupe*

**997** Arondissement Minéralogique, B.P. 448, Pointe-à-Pitre

*Guatemala*

**998** Instituto Geográfico Nacional, Avenida de las Américas 5–76, Zona 13, Guatemala City

*Guinea*

**999** Service Topographique et Géographique, B.P. 159, Conakry

*Guyana*

**1000** Topographic Division, Lands and Survey Department, Ministry of Agriculture, Vlissingen Road, Georgetown

**1001** Geological Survey Department, P.O. Box 789, Georgetown

*Haiti*

**1002** Service de Géodésie et de Cartographie, Département des Travaux Publiques, Boulevard Harry Truman, Cité de l'Exposition, Port-au-Prince

*Honduras*

**1003** Instituto Geográfico Nacional y Ministerio de Comunicaciones y Obras Públicas, Barrio La Bolsa, Comayagüela

*Hong Kong*

**1004** Crown Lands and Survey Office, Department of Lands, Survey and Town Planning, Murray Building, Hong Kong.

*Hungary*

**1005** Országos Földugyi es Térkepészeti Hivatal [National Office of Lands and Mapping], 1860 Kossuth Lajos tér, Budapest V.

**1006** Magyar Allami Foldtani Intezet [National Geological Institute], Nepstadion ut 14, Budapest XIV

**1007** Kartografiai Vallalat, Bosnyak tér 5, 1443 Budapest XIV

*Iceland*

**1008** Landmaelingar Islands, Laugavegur 178, Reykjavik

**1009** Sjømaelingar Islands, Seljavegi 32, P.O. 7094, Reykjavik

*India*

**1010** Map Division, Geological Survey of India, 27 Jawaharlal Nehru Road, Calcutta 13

**1011** Surveyor General's Office, Survey of India, P.O.B. 37, Hathibarkala, Dehra Dun

**1012** Naval Hydrographic Office, P.O.B. 75, Dehra Dun, 244 001

*Indonesia*

**1013** Badan Koordinasi Survey dan Pemetoan Nasional, Jalan Wahidin 1/11, Djakarta

**1014** Direktorat Tata Guna Tauoh, Jalan Seginghamangaya 2, 2nd floor, Kebayoran, Djakarta

**1015** Direktorate Jenderal Pertambangan Umum, Jalan M.H. Thamrin 1, Medan Merdelu Selatan 19, Djakarta

**1016** Jawatan Hidro-Oceanografi, Jalan Gunung Sahari 87, Djakarta

*Iran*

**1017** Iran National Cartographic Centre, B.P. 1844, Teheran

**1018** Iranian Geological Survey, K.H. Mehrabad, Teheran

*Iraq*

**1019** Department of Survey of Iraq, Baghdad

*Ireland*

**1020** Ordnance Survey, Phoenix Park, Dublin

**1021** Geological Survey, 14 Hume Street, Dublin 2

*Israel*

**1022** Survey of Israel, Ministry of Labour, 1 Lincoln Street, P.O. Box 14171, Tel Aviv 61140

**1023** Geological Survey, 30 Malkhei Israel Street, Jerusalem

**1024** Oceanographic and Limnological Research Ltd., Tel Shikmona, P.O. Box 1793, Haifa

*Italy*

**1025** Istituto Geografico Militare, Viale Filippo Strozze14, I-50100 Florence

**1026** Istituto Idrografico della Marina, Passo Osservatorio 4, I-16100 Genoa

*Ivory Coast*

**1027** Direction de l'Institut Géographique, Ministère des Travaux Publiques et des Transports, B.P. 20952, Abidjan

**1028** Service Géologique, B.P. 1368, Abidjan

*Jamaica*

**1029** Survey Department, P.O. Box 493, Kingston

**1030** Geological Survey, Hope Gardens, Kingston 6

*Japan*

**1031** Kokudo Chirin (Geographical Survey Institute), 24–13 Higoshiyam 3-Chome, Meguro-ku, Tokyo

**1032** Chishitsu Chosa-sho (Geological Survey), 8 Kawada-cho, Shinjuku-ku, Tokyo

**1033** Kaijōhoam Chō Suiro-Bu (Hydrographic Department), 5-chome, Tsukiji, Chuo-ku, Tokyo 104

*Jordan*

**1034** Department of Lands and Survey, P.O. Box 70, Amman

**1035** Geological Survey, Natural Resources Authority, P.O. Box 2220, Amman

*Kenya*

**1036** Public Map Office, Survey of Kenya, P.O. Box 30046, Nairobi

**1037** Mines and Geological Department, Ministry of Natural Resources, P.O. Box 30009, Nairobi

*Korea (South)*

**1038** Doe Han Min Gug, Su Ro Gug, C.P.O. Box 1578, 11 Leung Building, 2nd and 3rd Floors, 126–4 6a Chugma-Ro, Tungbu, Seoul

**1039** Geological Survey of Korea, 125 Namyoung-dong, Seoul

*Kuwait*

**1040** Chamber of Commerce and Industry, Chambers Building, Ali Salem Street, P.O. Box 775, Kuwait City

*Lebanon*

**1041** Service Technique du Cadastre, Grand Serail, Beirut

**1042** Etat-Major de l'Armée, Direction des Affaires Géographiques, Ministère de la Défense Nationale, Beirut

*Lesotho*

**1043** Lands and Survey Department, P.O. Box 876, Maseru

*Liberia*

**1044** Liberian Cartographic Service, Ministry of Lands and Mines, Monrovia

**1045** Liberian Geological Survey, Bureau of Natural Resources and Surveys, P.O. Box 9024, Monrovia

### Libya

**1046** Director of Surveying and Mapping, Ministry of Planning and Development, Bengazi

**1047** Geological Research and Mining Department, Industrial Research Centre, P.O. Box 3633, Tripoli

### Luxembourg

**1048** Administration du Cadastre et de la Topographie, 54 avenue Gaston Dieder, Luxembourg

**1049** Service Géologique, 13 rue J.P. Koenig, Luxembourg.

### Malagasy Republic (Madagascar)

**1050** Institut Cartographique de Madagasikara, 3 Lalàna Republika Malagasy Andohalo, B.P. 323, Tananarive

**1051** Centre de l'Institut Géographique National à Madagascar, B.P. 456, Tananarive

**1052** Service Géologique, Direction des Mines et de l'Energie, Ministère de l'Industrie et des Mines, B.P. 280, Tananarive

### Malawi

**1053** Map Sales Division, Survey Department, Ministry of Natural Resources, Lilongwe

**1054** Geological Survey, P.O. Box 27, Zomba

### Malaysia

**1055** Jabatanarah Penetaan Negara (Directorate of National Mapping), Jalan Gurney, Kuala Lumpur

**1056** Geological Survey of Malaysia, Scrivenor Road, P.O. Box 1015, Ipoh

### Mali

**1057** Institut National de Topographie, B.P.240, Bamako

### Martinique

**1058** Arrondissement Minéralogique de la Guyane, B.P. 468, Fort-de-France

### Mauritius

**1059** Ministry of Housing, Lands and Town and Country Planning, Port Louis

## Mexico

**1060** Dirección General de Geografía del Territorio Nacional, Avenida San Antonio Abad 124, México 8, D.F. México

**1061** Dirección General de Geografía y Meteorología, Avenida Observatorio 192, Colonia Tacubaya, México 18, D.F. México

**1062** Instituto de Geología, Biblioteca Universidad Nacional Autónoma de México, Ciudad Universitaria, Apdo. Postal 70-296, México 20, D.F. México

**1063** Dirección General de Oceanografía y Señalamiento Marítimo, Secretaria de Marina, México 1, D.F. México

## Morocco

**1064** Division de la Carte, Direction de la Conservation Foncière et des Travaux Topographiques, 31 Avenue Hassan Ier, Rabat

## Mozambique

**1065** Serviços Geográficos Cadastrais, C.P. 288, Maputo

**1066** Serviços de Geologia e Mines, C.P. 217, Can Phumo

## Namibia

**1067** Surveyor-General, Post Bag 13182, Windhoek 9100

## Nauru

**1068** Director of Lands and Surveys, Nauru Island, South Pacific

## Nepal

**1069** Survey Department, Topographical Survey Branch, Ministry of Land Reform, Kathmandu

## Netherlands

**1070** Topografische Dienst, Westvest 9, Postbus 145, Delft

**1071** Mapping Department, Rijks Geologische Dienst, Spaarne 17, Postbus 157, Haarlem

**1072** Afdeling Hydrografie van het Ministerie van Defensie-Marine, 171 Badhuisweg, The Hague.

## Netherlands Antilles

**1073** Bureau voor Kadaster, Willemstad, Curaçao

## New Zealand

**1074** Map Centre, Department of Lands and Survey, P.O. Box 6452, Te Aro, Wellington

**1075** New Zealand Geological Survey, Department of Scientific and Industrial Research, P.O. Box 30–368, Lower Hutt

**1076** Hydrographic Supplies Manager, Hydrographic Office, New Zealand Navy, P.O. Box 33–341, Akapuna, Auckland 9.

**1077** Soil Bureau, Department of Scientific and Industrial Research, Private Bag, Lower Hutt

*Nicaragua*

**1078** Ministerio de Fomento y Obras Públicas, Instituto Geográfico Nacional, Km 6 Carretera Norte, Apartado Postal 2110, Managua

**1079** Servicio Geológico Nacional, Ministerio de Economía, Apartado Postal 1347, Managua

*Niger*

**1080** Service Topographique et du Cadastre, Agence de Niamey, B.P. 250, Niamey

*Nigeria*

**1081** Federal Survey of Nigeria, Ministry of Works and Surveys, 5 Tafawa Balewa Square, P.M.B. 12596, Lagos

**1082** Geological Survey Division, Ministry of Mines and Power, P.M.B. 2007, Kaduna South, Northern Nigeria.

*Norway*

**1083** Norges Geografiske Oppmaling, St. Olavs Gt. 32, Oslo 1

**1084** Norges Geologiske Undersøkelse, Leif Erikssons Vei 39, N-7000 Trondheim

**1085** Norges Sjøkartwerk, Klubbgt. 1, Stavanger

*Pakistan*

**1086** Survey of Pakistan, Victoria Road, P.O. Box 3906, Karachi

**1087** Geological Survey of Pakistan, P.O. Box 15, Quetta

**1088** Hydrographic Directorate, Naval Headquarters, 11 Napier Barracks, Karachi

*Panama*

**1089** Instituto Geográfico Nacional 'Tommy Guardia', Apartado Postal 5267, Panama 5

**1090** Departamento de Recursos Minerales, Apartado Postal 1631, Panama

*Paraguay*

**1091** Instituto Geográfico Militar, Ministerio de Defensa Nacional, Avenida Artigas y via Ferrea, Casilla de Correo 316, Asunción

**1092** Dirección de la Producción Mineral, Tucari 271, Asunción

*Peru*

**1093** Instituto Geográfico Militar, Ministerio de Guerra, Apartado 2038, Lima

**1094** Servicio de Geología y Minería, Apartado 889, Lima

**1095** Dirección de Hidrografía y Navegación de la Marina, Sáenz Peña Cuadra 5ta, La Punta, Callao

*Philippines*

**1096** Bureau of Mines, Department of Agriculture and Natural Resources, P.O. Box 1595, Manila

**1097** Kawanihan ng Pagsukat sa Baybayin at Kalupaan, 421 Barrac San Nicolas, Manila D-405

*Poland*

**1098** Pánstwowe Przedsiebiorstwo Wydawnictw Kartograficznych (PPWK) [State Plant of Cartographic Editions], ul. Solic 18–20, 00–410 Warsaw

**1099** Instytut Geodezji i Kartografii [Institute of Geodesy and Cartography], ul. Jasna 2/4, 00–950 Warsaw

**1100** Pánstwowy Instytut Geologiczny [Geological Institute], ul. Rakowiecka 4, Warsaw

**1101** Biuro Hydrograficzne Marynarki Wojennej PRL, Gdynia 12

*Portugal*

**1102** Instituto Geográfico e Cadastral, Praca da Estrela, Lisbon 2

**1103** Serviços Geológicos de Portugal, Rue da Academia das Ciências 19–2, Lisbon

**1104** Instituto Hidrográfico, Rua das Trinas 49, Lisbon 2

*Romania*

**1105** Institutul de Geographie, Calea Victorie 126, Bucharest

**1106** Institutul de Geologie si Geografie, Str. Dr. Burghele 1, Bucharest 1

**1107** Comitetul de Stat al Geologiei, Calea Grivitei 64, Bucharest 12

*Rwanda*

**1108** Direction de la Cartographie, Ministère du Plan et Ressources Naturelles, Kigali

**1109** Services des Terres, Ministère de l'Agriculture et de l'Elevage, P.B. 621, Kigali

*Saudi Arabia*

**1110** Aerial Survey Department, Ministry of Petroleum, P.O. Box 247, Riyadh

**1111** Directorate General of Mineral Resources, P.O. Box 345, Jeddah

*Senegal*

**1112** Institut Géographique National, C.P. 4016, Dakar

**1113** Service des Mines et de la Géologie, B.P. 1238, Dakar

*Sierra Leone*

**1114** Survey and Lands Department, Freetown

**1115** Geological Department, Ministry of Lands and Mines, New England, Freetown

*Singapore*

**1116** Survey Department, 3rd Floor, National Development Building, Maxwell Road, Singapore

*Solomon Islands (British)*

**1117** Geological Survey, P.O. Box G-24, Honiara, Guadalcanal

*Somalia*

**1118** Geological Survey, Box 41, Hargeisa

*South Africa*

**1119** Director General of Surveys, P.O. Box 624, Pretoria 0001

**1120** Trigonometrical Survey Office, Rhodes Avenue, Mowbray, Cape Province

**1121** Director of Geological Survey, Private Bag X112, Pretoria 0001

**1122** The Hydrographer, Maritime Headquarters, Youngsfield, Kenwyn 7790

*Spain*

**1123** Distribución de Cartografía, Servicio Geográfico del Ejército, Ministerio de Ejército, Calle Prim No. 8, Madrid 4

**1124** Instituto Geográfico y Catastral, Calle del General Ibáñez de Ibero 3, Madrid

**1125** Instituto Geológico y Minero de España, Ríos Rosas 9, Madrid 3

**1126** Instituto Hidrográfico de la Marina, Tolosa Latour No. 1, Cadiz

*Sri Lanka*

**1127** Survey Department, Surveyor General's Office, P.O. Box 506, Colombo 5

**1128** Geological Survey, 48 Sri Jinaratana Road, Colombo 2

*Sudan*

**1129** Department of Surveys, Ministry of Defence, P.O. Box 306, Khartoum

**1130** Geological Survey Department, Ministry of Mining and Industry, P.O. Box 410, Khartoum

*Suriname*

**1131** Central Bureau for Aerial Mapping, Department of Development, P.O. Box 971, Paramaribo

**1132** Geologisch Mijnbouwkundige Dienst, Departement van Opbouw, Klein Wasserstraat, Paramaribo

*Swaziland*

**1133** Surveyor General, P.O. Box 58, Mbabane

**1134** Public Works Department, Ministry of Works, Power and Communication, P.O. Box 58, Mbabane.

**1135** Geological Survey and Mines Department, P.O. Box 9, Mbabane

*Sweden*

**1136** Generalstabens Litografiski Anstalt Forlag (GLA), Box 22069, S-104 22 Stockholm 22

**1137** Rikets Allmanna Kartverk, Hasselby Torg 20, Fack, S-162 10 Vallinby 1

**1138** Sveriges Geologiska Undersøkning, Stockholm

**1139** Sjøkartsverket, Sjøkarteavdelningen, Fack, S-601 01 Norrköping

*Switzerland*

**1140** Eidgenössische Landestopographie [Swiss Topographical Survey], Seftigenstrasse 264, CH-3084 Wabern, Berne

**1141** Schweizerische Geologische Kommission, Bernoullianum, CH-4000 Basel

*Syria*

**1142** Service Géographique de l'Armée, Damascus

**1143** Department of Geological Research and Mineral Prospecting, Jisr al Abiad, Damascus

*Taiwan*

**1144** Geological Survey of Taiwan, P.O. Box 31, Taipei

**1145** Chinese Naval Hydrographic and Oceanographic Office, P.O. Box 8505, Tse-Ying, Kao-Hsiung

*Thailand*

**1146** Royal Thai Survey Department, Geographic Mapping Division, Supreme Command Headquarters, Rachini Road, Bangkok 2

**1147** Land Development Department, Ministry of National Development, Rajdamnern Nok Avenue, Bangkok 2

**1148** Krom Uthoksat, Bangkok

*Togo*

**1149** Service Topographique, Ministère des Finances et de l'Economie, Lomé

**1150** Direction des Mines et de la Géologie, Ministère des Travaux Publics, Mines, Transports, des Postes et Télécommunications, B.P. 356, Lomé

*Tonga*

**1151** Superintendent of Lands and Surveys, Nuku' Alofa, Tongatapu Island

*Tunisia*

**1152** Division Topographique, Secrétariat d'Etat aux Travaux Publics, rue de Jordanie 13, Tunis

**1153** Geological Survey, 95 avenue Mohamed V, Tunis.

*Turkey*

**1154** Mapping Service, Ministry of National Defence, Harita Genel Müdürlügü, Cebeci, Ankara

**1155** Seyir Hidrografi ve Osinografi Dairesi Baskanligi, Cubuklu, Istanbul

*Uganda*

**1156** Map Sales Office, Department of Lands and Surveys, 15 Obote Avenue, P.O. Box 361, Kampala

**1157** Geological Survey and Mines Department, P.O. Box 9, Entebbe

*United Kingdom*

**1158** Ordnance Survey, Romsey Road, Maybush, Southampton SO9 4DH

**1159** Ordnance Survey of Northern Ireland, 83 Ladas Drive, Belfast BT6 9F1

**1160** British Geological Survey, Exhibition Road, South Kensington, London SW7 2DE

**1161** Admiralty Hydrographic Service, Creechbarrow House, Taunton, Somerset

*United States of America*

**1162** Geological Survey, National Center, 12201 Sunrise Valley Drive, Reston, Virginia 22092

**1163** National Ocean Survey, National Oceanic and Atmospheric Administration, 6010 Executive Boulevard, Rockville, Maryland 20852

**1164** National Ocean Survey, Lake Survey Center, 630 Federal Building, Detroit, Michigan 48226.

**1165** Defense Mapping Agency Hydrographic Center, Washington, D.C. 20390

**1166** Defense Mapping Agency Aerospace Center, St. Louis Air Force Station, St. Louis, Missouri 63118

*Upper Volta*

**1169** Servicio Geográfico Militar, Avenida 8 de Octubre 3255, Montevideo

**1168** Direction de la Géologie et des Mines, B.P. 601, Ouagadougou

*Uruguay*

**1169** Servicio Geográfico Militar, Avenida 8 de Octubre 3255, Montevideo

**1170** Instituto Geológico del Uruguay, Calle Jalio Herrara y Obes 1239, Montevideo

**1171** Servicio de Oceanografía e Hidrografía de la Armada, Capurro 980, Casilla de Correo 1380, Montevideo

*U.S.S.R.*

**1172** Chief Administration of Geodesy and Cartography, Council of Ministers of the U.S.S.R., Krzhizhanovskogo 14, Korpus 2, Moscow V-218

**1173** Hydrograficheskoe Upravlenie, Ministerstvo Oboroni, 8, 11 Liniya, Leningrad V-34

*Venezuela*

**1174** Dirección de Cartográfica Nacional, Caracas

**1175** Dirección de Geología, Ministerio de Minas e Hidrocarburos, Torre Norte, Piso 19, Caracas

**1176** Comandancia General de la Marina, Dirección de Hidrografía y Navegación, AP NC 6745, Caracas

*Yugoslavia*

**1177** Geo-Karta Institute, Belgrade

**1178** Jugoslavenski Leksikografski Zavod, Zagreb

**1179** Hidrografski Institut Jugoslavenske Ratne Morrarice, Split

*Zaire*

**1180** Institut Géographique de Zaïre, 106 boulevard du 30 Juin, B.P. 3086, Kinshasa

**1181** Service Géologique, Direction des Etudes et Recherches du Département des Mines, B.P. 3149A, Kinshasa

*Zambia*

**1182** Geological Survey Department, Ministry of Mines, P.O. Box 50135, Ridgeway, Lusaka

**1183** Ministry of Lands and Natural Resources, Surveys Department, P.O. Box 50397, Lusaka

*Zimbabwe*

**1184** Department of the Surveyor General, Ministry of Lands, P.O. Box 8099, Harare

## Selective List of Commercial Publishers

The number of firms that produce maps, atlases, globes and other cartographic material commercially is a formidable one and the following list is highly selective. Much of the material produced is directed to the educational or leisure fields and many of the products are 'derived' maps rather than those made from original survey. Numerous establishments, however, do carry out surveys on their own account. Enquiries for information are likely to produce an illustrated catalogue of products or possibly brochures describing the type of work carried out. Tourist organizations are excluded from the list, though some produce excellent maps.

## Argentina
**1185** Peuser SA, San Martin 200, Buenos Aires

## Australia
**1186** Jacaranda Wiley Ltd., 65 Park Road, Milton, Queensland 4064

**1187** Rigby Educational, 594 St. Kilda Road, Melbourne, Victoria 3004

**1188** Universal Business Directories Pty. Ltd., 64 Talavera Road, North Ryde, New South Wales

## Austria
**1189** H. Fleischmann KG (Kompass), Deffregger Str. 36, A-6020 Innsbruck

**1190** Freytag-Berndt u. Artaria KG, Schottenfeldgasse 62, A-1071 Wien III

**1191** O. Müller Verlag, Ernest-Thun Str. II, Pf. 167, A-5021 Salzburg

**1192** Österreichischer Alpen Verein, A-6020 Innsbruck

## Belgium
**1193** R. de Rouck, rue de la Ruche 3, Brussels

## Canada
**1194** Canadian Cartographics Ltd., 508 Clark Road, Coquitlam, British Columbia

**1195** Gage Publishing Ltd., 164 Commander Blvd., Agincourt, Ontario M1S 3C7

## Czechoslovakia
**1196** Kartograficke Nakladatelství, Kostelní 42, CS-120 20 Prague 7

**1197** Slovenská Kartografia NP, Pekná Cesta 17, Bratislava-Krasňany

## Denmark
**1198** Gyldendal Forlag, Klareboderne 3, Pb.11, DK-1001 Copenhagen K

**1199** Kraks, Nytorv 17, DK-1450 Copenhagen K

**1200** Politikens Forlag, Vestergade 26, DK-1456 Copenhagen K

*France*
**1201** Armand Colin, 103 blvd. St. Michel, F-75005 Paris

**1202** Blondel La Rougery, 7 rue St. Lazare, Paris 9

**1203** Cartes Taride, 2 bis Place du Puits de l'Ermite, F-75005 Paris

**1204** Editions Bordas, 11 rue Gossin, F-92543 Montrouge

**1205** Editions Ponchat, 7 rue Theodore de Banville, Paris 17

**1206** Editions MDI, Parc des Dix Arpents D113, F-78630 Orgeval

**1207** Hachette, 79 blvd. St. Germain, F-75006 Paris

**1208** Lib. Larousse, 17 rue Montparnasse, F-75006 Paris

**1209** Plans-Guides Blay, 26 rue Feydeau, Paris 2

**1210** Service de Tourisme Michelin, 46 avenue de Breteuil, F-75341 Paris

*German Democratic Republic*220
**1211** VEB Hermann Haack, Geographisch-Kartographische Anstalt, Justus Perthes Str. 3–9, Pf. 274, DDR-Gotha

*Federal Republic of Germany*
**1212** Columbus Verlag, Columbus-Haus, Pf. 1180, D-7056 Weinstadt 1

**1213** Falk Verlag, Burchardstr. 8, D-2000 Hamburg 1

**1214** Fritsch Landkartenverlag, Ludwig Str. 7, Pf. 751, D-8000 Munich 22

**1215** Kartographische Inst. Bertelsmann GmbH, Pf. 5555, D-4830 Gütersloh 1

**1216** Mairs Geographischer Verlag, Marco Polo Str. 1, D-7302 Ostfildern 1

**1217** Justus Perthes, Holzhofallee 36b, Pf. 649, D-6100 Darmstadt 2

**1218** Polyglott Verlag, Pf. 40 11 20, D-8000 Munich 40.

**1219** Ravenstein Verlag, Wielandstr. 31–35, D-6000 Frankfurt 1

**1220** RV (Reise- und Verkehrsverlag GmbH), Schockenriedstr. 40a, Pf. 80 08 63, D-7000 Stuttgart 80

**1221** Georg Westermann Verlag, Georg-Westermann-Allee 66, Pf. 3320, D-3300 Braunschweig

*Greece*

**1222** Alma, Menandrou 68, Athens

**1223** D-B. Loukopolous, Stoa Nikoloudi 10, Athens

*India*

**1224** Dipti Publications, Sri Aurobindo Ashram, Pondicherry 2

**1225** Map and Atlas Publications Pvt. Ltd., 331–333 Thambu Chetty Street, Madras 600 001

*Indonesia*

**1226** Pembina, Papandanjan 41, Djakarta

*Iran*

**1227** Sahab Geographic and Drafting Inst., P.O.B. 263, Teheran

*Israel*

**1228** Amir Publishing, 5 Engel Street, Tel Aviv

**1229** Carta, Yad Harutzim Street, Jerusalem 91020

*Italy*

**1230** Istituto Geografico de Agostini, Via Giovanna da Verrazano 15, CP 157, I-28100 Novara

**1231** Studio FMB, Bologna, SNC—Via A. Costa 142/2—Rastignano (BO)

**1232** Touring Club Italiano, Corsa Italia 10, I-20122 Milan

*Japan*

**1233** Teikoku-Shoin Co. Ltd., 3–29 Jimbo-Cho, Kanda, Chiyoda-Ku, Tokyo 101

**1234** Tokyo-Chizu Co., 3–4 Jimbo-Cho, Chiyoda-Ku, Tokyo

*Mexico*

**1235** Patria SA, Av. Uruguay 25, Apdo. 784, México 1, D.F.

*Netherlands*

**1236** ANWB, Wassenaarseweg 220, The Hague

**1237** Born NV, Haspelsstr. 2–4, Pb. 8060, Amsterdam

**1238** Suurland's, Dommelstr. 36, Eindhoven

*New Zealand*

**1239** A. H. & H. W. Reed, 182 Wakefield St., Wellington

*Norway*

**1240** J. W. Cappelens Forlag, Kirkegt. 15, Oslo 1

**1241** Rutebook for Norge, P.B. 153 Sentrum, Oslo 1

*South Africa*

**1242** Juta & Co. Ltd., Mercury Crescent, Hillstar Industrial Township, C.P. 7700

**1243** J. L. van Schaik Pty. Ltd., P.O.B. 724, Pretoria 0001

*Spain*

**1244** Aguilar SA, Juan Bravo 38, Madrid 6

**1245** Alpina, Navarra 58, Granollers, Barcelona

**1246** Editorial Teide SA, Calle Viladomat 291, Barcelona 15

*Sweden*

**1247** Esselte Map Service, Garvargatan 9, P.O. Box 22069, S-104 22 Stockholm

*Switzerland*

**1248** Hallwag AG, Nordring 4, CH-3000 Berne

**1249** Kümmerly & Frey AG, Hallerstrasse 6–12, CH-3001 Berne

**1250** Orell Füssli Verlag, Nüschelerstr. 22, CH-8022 Zurich

*United Kingdom*

**1251** AA Publications, Fanum House, Basingstoke, Hampshire RG21 2EA

**1252** G. I. Barnett & Son Ltd., Rippleside Commercial Estate, Ripple Road, Barking, Essex

**1253** J. Bartholomew & Son Ltd., 12 Duncan Street, Edinburgh EH9 1TA

**1254** Clyde Surveys, Reform Road, Maidenhead, Berkshire

**1255** Collins-Longman, Burnt Mill, Harlow, Essex CM20 2JE

**1256** Estate Publications, Rushton, Smarden, Ashford, Kent

**1257** Forward Publicity Ltd., Falcon House, 20–2 Belmont Road, Wallington, Surrey SM6 8TA

**1258** Geographers A–Z Map Co. Ltd., Sevenoaks, Kent.

**1259** Geographia Ltd., 63 Fleet Street, London EC4

**1260** Geoprojects (UK) Ltd., Newtown Road, Henley-on-Thames, Berkshire.

**1261** Chas. E. Goad Ltd., 18A Salisbury Square, Old Hatfield, Hertfordshire AL9 5BE

**1262** Imray, Laurie, Norie & Wilson Ltd., Wych House, The Broadway, St. Ives, Cambridgeshire PE17 4BT

**1263** Macmillan Publishers Ltd., 4 Little Essex Street, London WC2R 3LF

**1264** Map Productions Ltd., 27A Floral Street, London WC2E 9LP

**1265** Oxford University Press, Walton Street, Oxford OX2 6DP

**1266** Pergamon Press Ltd., Headington Hill Hall, Oxford OX3 0BW

**1267** Quail Map Co., 31 Lincoln Road, Exeter, Devon EX4 2DZ

**1268** Thomas Nelson & Sons Ltd., Nelson House, Mayfield Road, Walton-on-Thames KT12 5PL

*United States of America*

**1269** American Map Co., 46–35 54th Road, Maspath, New York 11378

**1270** Arrow Publishing Co. Inc., P.O.B. 466, Astor Street, Boston, Massachusetts 02123

**1271** Champion Map Co., 715 Norwell Place, Charlotte, North Carolina 28205

**1272** Denoyer-Geppert Co., 5235 W. Ravenswood Avenue, Chicago, Illinois 60640

**1273** Dolph Map Co. Inc., 430 N. Federal Highway, Fort Lauderdale, Florida 33301

**1274** General Drafting Co Ltd., Canfield Road, Convent Sta., New Jersey 07961

**1275** G. F. Cram Co. Inc., 301 South La Salle Street, Indianapolis, Indiana 46206

**1276** Geo Map Co., P.O.B. 30008, Dallas, Texas 75230

**1277** H. M. Gousha Publications, P.O.B. 6227, San Jose, California 95114

**1278** Hagstrom Co. Inc., 450 W. 33rd Street, New York, NY 10001

**1279** Hammond Inc., 515 Valley Street, Maplewood, New Jersey 07040

**1280** Hearne Bros., 25th Floor, First National Building, Detroit, Michigan 48226

**1281** Historic Urban Plans, P.O.B. 276, Ithaca, New York 14850

**1282** National Geographic Society, 17th & M Streets, N.W. Washington, D.C. 20036

**1283** A. J. Nystrom & Co., 3333 Elston Avenue, Chicago, Illinois 60618

**1284** Rand McNally & Co., P.O.B. 7600, Chicago, Illinois 60680

**1285** Thomas Bros. Maps, 1326 South Broadway, Los Angeles, California 90015

*Zimbabwe*

**1286** The College Press (PVT) Ltd., P.O.B. 3041, Harare

## Selected Publishers and Distributors of Early Map Facsimiles

**1287** Arti Grafiche Ricardi, Via Cortina d'Ampezzo, 10–20139 Milan, Italy. (Colour facsimiles of European maps.)

**1288** John Bartholomew & Son Ltd., 12 Duncan Street, Edinburgh EH9 1TA, Scotland. (Colour facsimiles of seventeenth-century maps, mainly by Blaeu.)

**1289** British Library, Reference Division, Great Russell Street, London WC1B 3DG, England. (Various colour and monochrome facsimiles including Saxton's county maps; Matthew Paris's maps; Boazio's map of Ireland; six early printed maps.)

**1290** Cambrian Distributors, Holywell, Clwyd, Wales. (Seventeenth-century maps.)

**1291** Culture et Civilisation, 115 Avenue Gabriel Lebon, Brussels, Belgium. (Facsimile of Mercator's *Atlas*.)

**1292** David & Charles, Brunel House, Forde Road, Newton Abbot, Devon TQ12 4PU, England. (Reprint of the Old Series of one inch to one mile Ordnance Survey maps; Bowen's *Royal English Atlas*; Camden's *Britannia*, 1695 edition; Morden's county maps from *Britannia*, 1695.)

**1293** Edition Leipzig, P.O.B. 340, 701 Leipzig, German Democratic Republic. (*Africa in Maps.*)

**1294** Editio Totius Mundi, Gussenbauergasse 5/9, A-1090 Vienna, Austria. (Extensive range of facsimiles by many mapmakers together with town plans and views by Merian, Schedel, Braun and Hogenberg *et al.*)

**1295** EP Group of Companies, Reprint Division, East Ardsley, West Yorkshire, England. (Speed's county maps of Wales, 1676.)

**1296** Essex Record Office, Chelmsford, Essex CM 1LX, England. (Facsimiles of Essex county maps).

**1297** Evans, Olwen Caradoc, Perllan Caradoc, Conwy, Gwynedd, Wales. (Maps and charts of Wales.)

**1298** Generalstabens Litografiska Anstalts Förlag, Vasagatan 16, Stockholm, Sweden. (Olaus Magnus's map of Scandinavia, 1539; plans of Stockholm.)

**1299** Alan Godfrey, Dunston, Gateshead, Tyne & Wear, England. (Reduced facsimiles of early Ordnance Survey large-scale plans.)

**1300** Frank Graham, 6 Queens Terrace, Jesmond, Newcastle-upon-Tyne, England. (Bowen's *Britannia Depicta*, 1720; *Bickham's British Monarchy*, 1748, etc.)

**1301** Hermann Haack VEB, Geographisch-Kartographische Anstalt, Gotha/Leipzig, German Democratic Republic. (*Schöne Alte Karten*, portfolio of early maps; early map calendars.)

**1302** International Antiquariaat, Keizersgracht 61, Amsterdam, The Netherlands. (Plancius World Map, 1604.)

**1303** Israel, N., Keizersgracht 539, Amsterdam, The Netherlands. (Facsimile editions of various rare atlases.)

**1304** Istituto Poligrafico dello Stato, Piazza Verdi 10, Rome, Italy. (Fra Mauro's world map, 1458.)

**1305** London Topographical Society, 9 Riverscourt Road, London W6, England. (Maps and plans of London.)

**1306** Harry Margary, Lympne Castle, Kent. (Eighteenth-century English county maps; sets of county maps by counties; American maps.)

**1307** Redpath, Campbell and Partners Ltd., Cranham, Gloucestershire, England. (Modern county maps.)

**1308** Philip Riden, 14 The Croft, Old Headington, Oxford, England. (Eighteenth-century county maps.)

**1309** Carlos Sanz, Bibliotheca Americana Vetustissima, Velasquez 6, Madrid 1, Spain. (Important collection of world maps.)

**1310** Schuler Verlag, Stuttgart, Federal Republic of Germany. (Early world maps; large range of regional maps; marine charts by Waghenaer).

**1311** Sidgwick & Jackson Ltd., 1 Tavistock Chambers, Bloomsbury Way, London WC1A 2SG, England. (Panoramas of London.)

**1312** Stevens, Brian, Historic Prints, 11A Mayhill, Monmouth. (Nineteenth-century county maps and town plans by Moule, Cole and Roper, etc.)

**1313** Theatrum Orbis Terrarum, Singel 157, Amsterdam, The Netherlands. (Important facsimile editions of many rare atlases.)

**1314** Charles Traylen, Castle House, 49–50 Quarry Street, Guildford, Surrey, England. (Speed's county maps; Buck's views of English towns.)

**1315** University of Exeter, The Registry, Northcote House, Queen's Drive, Exeter, Devon, England. (Benjamin Donn's map of Devon; John Norden's Ms. maps of Cornwall.)

**1316** Verlag Walter Uhl, P.O. Box 1, D-7091 Unterscheidheim, Federal Republic of Germany. (Portfolio of important maps in cartographic history; town views.)

**1317** West of England Press, 1 West Street, Tavistock, Devon. (1520 edition of Ptolemy's *Geographia*.)

## Some American Publishers of Facsimile Maps

**1318** American Elsevier Publishing Co. Inc., Vanderbilt Avenue, New York, NY 10017. (Facsimiles of important early atlases.)

**1319** American Heritage Publishing Co., 551 Fifth Avenue, New York, NY 10017. (Sixteenth-, seventeenth- and eighteenth-century maps.)

**1320** Art Fair Inc., 112 Fourth Avenue, New York. (Seventeenth- and eighteenth-century maps of North America.)

**1321** Family Circle, Department 936, Box 1379, Grand Central Station, New York, NY 10017. (Braun and Hogenberg town plans and views.)

**1322** Historic Plans, Box 276, Ithaca, New York. (Important range of town plans including many of North American cities.)

**1323** John Carter Brown Library, Providence, Rhode Island 02912. (Early maps of North America.)

**1324** Kentucky University Libraries, Lexington, Kentucky 40506. (Facsimiles of maps of Kentucky.)

**1325** Library of Congress, Information Office, Navy Yard Annex, Building 159, Washington, DC 20540. (Facsimile of John Smith's Virginia map).

**1326** North Carolina State Department of Archives and History, Box 1881, Raleigh, North Carolina 77602. (Sets of facsimiles of North Carolina maps.)

**1327** Parade, P.O. Box 589, Westport, Connecticut 06881. (Sixteenth- and seventeenth-century maps.)

**1328** Tudor Publishing Co., 221 Park Avenue South, New York, NY 10003. (Various early map facsimiles.)

## Associations and Societies

Many associations and societies devoted to cartography and related disciplines are now in existence throughout the world. Some publish excellent journals, such as The British Cartographic Society's *The Cartographic Journal* or the Society of University Cartographers' *SUC Bulletin*, that may be obtainable on subscription by non-members. Many societies publish newsletters for the benefit of their own members. Symposia, summer schools and lecture meetings are regularly organized and it is sometimes possible for non-members to attend.

### International

**1329** International Cartographic Association. Hon. Secretary: Olof Hedbom, Flottbrovögen 16, S-112 64 Stockholm, Sweden. (A list of member countries is provided on pp. 46–52 of *Orbis Geographicus* [673]. The same publication also lists the Commissions and Working Groups of ICA on pp. 53–56. Unfortunately these lists are very out of date.)

**1330** International Society for the History of Cartography. Secretary: G. R. P. Lawrence, Department of Geography, King's College, Strand, London WC2, England.

**1331** International Map Collectors' Society, S. Luck, 83 Marylebone High Street, London W1M 4AL, England.

**1332** International Society of Map Stamp Collectors. R. J. Richmond, 40 Clinton Street, 5M Brooklyn, New York 11201.

**1333** International Federation of Library Associations, Geography and Maps Section. Secretary: David Carrington, 5422 Marsh Hawk Way, Columbia, Maryland 21045

# National and regional societies

*Australia*

**1334** The Australian Institute of Cartographers, P.O. Box 1292, Canberra City, ACT 2601.
Regional divisions of the above:

**1335** *Victoria*: Kelvin Hall, 55 Collins Place, Melbourne.

**1336** *New South Wales*; Box 4365, G.P.O., Sydney.

**1337** *Australian Capital Territory*: 19 Riley Street, Turney, ACT

**1338** *Tasmania*: Box 705F, G.P.O., Hobart

**1339** *South Australia*: P.O. Box 1922, Port Adelaide

**1340** Institute of Cartographers of Western Australia, P.O. Box H.592, G.P.O., Perth, Western Australia

**1341** Australian Map Circle, P.O. Box E133, Queen Victoria Terrace, Canberra, ACT 2600

*Austria*

**1342** Kartographische Kommission der Österreichischen Geographischer Gesselschaft, Universitätstrasse 7, A-1010 Vienna. (General Secretary: Dr. Ingrid Kretschmer.)

*Belgium*

**1343** Société Belge de Cartographie/Belgische Vereniging voor Kartografie, c/o Institut Geographique Militaire/Militair Geografisch Instituut, 13 Abbaye de la Cambre, Brussels 5

*Brazil*

**1344** Sociedade Brasiliera de Cartografia, Rua Mexico 41, Gr. 706 entro, Rio de Janeiro.

*Canada*

**1345** Canadian Cartographic Association, Department of Geography, University of Ottawa, Ottawa, Ontario K1N 6N5

**1346** Canadian Institute of Surveying, Cartographic Committee, 157 McHead Street, Ottawa-4, Ontario

**1347** Ontario Institute of Chartered Cartographers, Department of Geography, York University, Toronto, Ontario

**1348** Association of Canadian Map Libraries, National Map Collection, Public Archives of Canada, Ottawa, Ontario K1A 0N3

**1349** Association Québecoise de Cartographie, CP 8684 Ste. Foy, G1V 4N6

## China
**1350** Chinese Society of Geodetics and Cartography, Beijing

## Czechoslovakia
**1351** Comité de Cartographie de la Societé Tchécoslovaque Scientifico-Technique, Široka 5, 11000 Prague 1

## Finland
**1352** Suomen Kartografinen Seura R.Y. [Cartographic Society of Finland], Kirkkokatu 3, SF-00170 Helsinki 17

## France
**1353** Société Cartographique de France, 48 rue de Charenton, F-75012 Paris.

**1354** Association Française pour le Développement de l'Expression Cartographique, 25 Domaine Defontaine, Achères-la-Forêt

**1355** Société Française d'Études et de Réalisations Cartographiques, 5 rue Papillon, F-75009 Paris

**1356** Société Graphique et Cartographique, 83 rue Blomet, F-75015 Paris

## German Democratic Republic
**1357** Sektion Kartographie in der Geographischen Gesellschaft der Deutschen Demokratischen Republic [Cartography Section of the G.D.R. Geographical Society], Dimitroffplatz 1, DDR-701 Leipzig

## Federal Republic of Germany
**1358** Deutsche Gesellschaft für Kartographie E.V., Klüsenerskamp 10, D-4600 Dortmund

## Hungary
**1359** Geodézial és Kartográfiai Egysület [Geodetic and Cartographical Association], Anker-köz 1, H-1061 Budapest VI

## India
**1360** Indian National Cartographic Association, Hyderabad

## Iran
**1361** National Cartographic Centre, P.O. Box 1844, Shayad Square, Teheran

*Ireland*

**1362** Irish Society of Surveying and Photogrammetry. Secretary: Survey House, 67 Lower Baggot Street, Dublin 2

*Israel*

**1363** Israeli Cartographic Society, c/o Survey of Israel, P.O. Box 14171, Tel Aviv 61140

*Italy*

**1364** Associazione Italiana di Cartografia, c/o Istituto Geografico Militare, Via Cesare Bottisti 10, Florence

*Japan*

**1365** Japan Cartographers' Association, Japan Map Centre, Kudan Ponpian Building, 8.8, 4-chome, Kudan-minami Chiyoda-ku, Tokyo 102

*Netherlands*

**1366** Kartografische Sectie van het Koninklijk Nederlands Aardrijkskundig Genootschap [Cartographic Section of the Royal Netherlands Geographical Society], St. Annastraat 418, Nijmegen

**1367** Nederlands Kartografische Vereeniging [Netherlands Cartographic Society], Sterkenburgerlaan 32, NL-3941 BD Doorn.

*New Zealand*

**1368** New Zealand Cartographic Society Inc., P.O. Box 9331, Courtnenay Place, Wellington

**1369** New Zealand Map Keepers' Circle. Secretary: P.O. Box 10.179, Terrace Post Office, Wellington

*Norway*

**1370** Norsk Kartografisk Forening [Norwegian Cartographical Society], c/o Widerøes Flyveselskap, Kristian Augusts Gate 9, Oslo

*South Africa*

**1371** South African Society for Photogrammetry, Remote Sensing and Cartography. Secretary: P.O. Box 69, Newlands 7725

**1372** Southern Africa Map Collectors' Association, University Library, University of Natal, P.O. Box 375, Pietermaritzburg 3200

*Spain*

**1373** Asociación Española para el Progreso de las Ciencias, Seminario de Cartografía [Spanish Association for the Advancement of Sciences, Cartographic Seminar], Valverde 24, Madrid

## Sweden

**1374** Kartografiska Sälskapet [Swedish Cartographic Society], c/o Esselte Map Service, Fack, S-105 31 Stockholm

## Switzerland

**1375** Schweizerische Gesellschaft für Kartographie [Swiss Cartographic Society], Institut für Kartographie, ETM-Hönggerberg, CH-8093 Zürich

## United Kingdom

**1376** The British Cartographic Society. Hon. Secretary: P. E. Sorrell, Department of Land Surveying, North East London Polytechnic, Longbridge Road, Dagenham, Essex RM8 2AS. Within BCS there are two specialized groups, viz. the Map Curators' Group and the Digital Cartography Group

**1377** Society of University Cartographers. Hon. Secretary: Huw Dobson, Department of Maritime Studies, UWIST, Aberconway Building, Colum Drive, Cardiff CF1 3EU

**1378** The Charles Close Society for the Study of Ordnance Survey Maps. Hon. Secretary: Brian Garvan, 66 Chalklands, Bourne End, Buckinghamshire SL8 5TJ.

## United States of America

**1379** American Cartographic Association, 210 Little Falls St., Falls Church, Virginia 22046

**1380** American Congress on Surveying and Mapping, 210 Little Falls St., Falls Church, Virginia 22046

**1381** American Library Association Maps and Geography Round Table. J. Walsh, Maps/Documents Librarian, Coe Library, University of Wyoming, University Station, Box 3334, Laramie, Wyoming 8207

**1382** California Map Society. N. Diaz, Bunche Hall, University of California, Los Angeles

**1383** Cartographic Users' Advisory Council. C. A. Seavey, General Library, University of New Mexico, Albuquerque, New Mexico 87131

**1384** Cartographic Speciality Group, Association of American Geographers. 1710 16th St., Washington, D.C. 20009

**1385** Geography and Map Division, Special Libraries Association. R. S. Green, Map Collection, University of Iowa Libraries, Iowa City, Iowa 52242

**1386** Geoscience Information Society, 4220 King Street, Alexandria, Virginia 22302

**1387** Map On-Line Users' Group, 235 Park Avenue South, New York, NY 10003

**1388** Map Society of the Delaware Valley. R. Hornick, 2401 Penn Avenue, 18B 30 Philadelphia, Pennsylvania 19130

**1389** New York Map Society, Map Division, New York Public Libraries, 42nd Street/5th Avenue, New York, NY 10018

**1390** Washington Map Society, 3051 Idaho Avenue, MW222, Washington, D.C. 20016

**1391** Western Association of Map Libraries. S. Mullin, 456 Alcatraz Avenue, Oakland, California 94605

## Remote Sensing and Photogrammetric Societies

**1392** International Society of Photogrammetry and Remote Sensing, United States Geological Survey, Mailstop 516, Reston, Virginia 22092

**1393** Remote Sensing Society. Secretary, c/o Geography Department, University of Reading, Reading, Berkshire RG6 2AU, England.

**1394** The Photogrammetric Society. Secretary, Department of Photogrammetry and Surveying, University College London, Gower Street, London WC1E 6BT, England

**1395** Irish Society of Surveying and Photogrammetry. Secretary: Survey House, 67 Lower Baggot Street, Dublin 2, Ireland

## Exploration and Discovery Societies

**1396** Society for the History of Discoveries. Secretary: Barbara B. McCorkle, 45 Mill Rock Road, Hamden, Connecticut 06511

**1397** The Hakluyt Society (for all who are interested in the literature of travel and in the history of geographical science and discovery). The Administrative Assistant, The Hakluyt Society, c/o The British Library, Great Russell Street, London WC1B 3DG, England

# Index

Numbers in brackets refer to numbered items in Parts II–IV, not to page numbers.